COSMIC IMPACT

COSMIC IMPACT

Understanding the Threat to Earth from Asteroids and Comets

ANDREW MAY

ICON

Published in the UK in 2019
by Icon Books Ltd, Omnibus Business Centre,
39–41 North Road, London N7 9DP
email: info@iconbooks.com
www.iconbooks.com

Sold in the UK, Europe and Asia
by Faber & Faber Ltd, Bloomsbury House,
74–77 Great Russell Street,
London WC1B 3DA or their agents

Distributed in the UK, Europe and Asia
by Grantham Book Services,
Trent Road, Grantham NG31 7XQ

Distributed in the USA
by Publishers Group West,
1700 Fourth Street, Berkeley, CA 94710

Distributed in Australia and New Zealand
by Allen & Unwin Pty Ltd,
PO Box 8500, 83 Alexander Street,
Crows Nest, NSW 2065

Distributed in South Africa
by Jonathan Ball, Office B4, The District,
41 Sir Lowry Road, Woodstock 7925

Distributed in India by Penguin Books India,
7th Floor, Infinity Tower – C, DLF Cyber City,
Gurgaon 122002, Haryana

Distributed in Canada by Publishers Group Canada,
76 Stafford Street, Unit 300
Toronto, Ontario M6J 2S1

ISBN: 978-178578-493-4

Typeset in Iowan by Marie Doherty

Printed and bound in Great Britain
by Clays Ltd, Elcograf S.p.A.

ABOUT THE AUTHOR

Andrew May is a freelance writer and science consultant. He has written on subjects as diverse as the physical sciences, military technology, British history and the paranormal. His recent books include pocket-sized biographies of Newton and Einstein, an eye-opening study of the relationship between pseudoscience and science fiction, and *Destination Mars*, in the Hot Science series. He lives in Somerset.

CONTENTS

ASTEROID APOCALYPSE 1

In the popular imagination, an asteroid or comet impact is one of the top 'end of the world' scenarios. It's a relative newcomer to the field, only having entered mainstream consciousness in the last decade of the 20th century – thanks in large part to the Hollywood films *Armageddon* and *Deep Impact* – but it took hold quickly and has stayed with us ever since.

If the gold standard of cliché-dom is to be satirised in *The Simpsons* cartoon series, then cosmic impact makes the grade. It's right there in episode 492 from 2011. Bart says: 'No matter what we do, an asteroid is going to wipe us out. So we should party hard and wreck the place!' – to which Homer replies: 'Yeah, why should the asteroid have all the fun?'

Popular culture aside, impacts from outer space really do pose a serious, ever-present threat. There are thousands of asteroids travelling on orbits that cross our own – and any of these could, in theory, end up in exactly the same place as the Earth at the same time. The same is true of many comets

– and new comets are falling into the inner Solar System all the time. From a cosmic perspective, impacts aren't a rarity – they're business as usual.

It's happened before …

Let's start by looking at the Moon. Its appearance has been shaped almost exclusively by impact events. Aside from the thousands of obvious impact craters, virtually all its other visible features – such as mountain ranges and the large dark areas that are misnamed 'seas' – were caused by impacts too. So why has the Moon suffered from cosmic collisions so much more than the Earth?

The answer is that it hasn't. The Earth has suffered just as much as the Moon, but various atmospheric and geological processes – not to mention the water that covers almost three-quarters of the planet – conspire to cover up the evidence over the course of centuries. Today, there's only one easily recognisable impact crater on Earth, the kilometre-wide Meteor crater (also known as Barringer crater) in Arizona. It's relatively fresh-looking (see page 60) because it's less than 50,000 years old – the blink of an eye in the planet's 4.5-billion-year history. However, if you know the signs to look out for, the Earth has plenty of other impact craters. They're generally older, and sometimes larger – much larger.

Chicxulub crater, for example, sprawls for almost 200 km across the northern tip of Mexico's Yucatan peninsula and out into the Gulf of Mexico. It's the relic of a catastrophic collision that occurred 66 million years ago, when a rocky object about 10 km across plummeted into the Earth from outer space. Now that's a big chunk of rock, and it would obviously have caused tremendous destruction on a local scale. On the other hand, it's not really that big compared to a planet that's almost 13,000 km in diameter. So the Earth as a whole would have hardly noticed the impact, right?

Wrong. The Chicxulub impactor was travelling at around 20 km/s (kilometres per second), which translates to an enormous amount of kinetic energy. When it hit, all that energy was transferred to the Earth in the form of a huge explosion. The biggest explosion that most people can visualise is the one that destroyed the city of Hiroshima in 1945. So let's try to imagine something five billion times worse than that, all concentrated in a single instant and at a single point on the Yucatan peninsula. That was Chicxulub.

The impact produced a huge cloud of dust and ash which enshrouded the planet and changed the global climate for centuries to come. The result was the extinction of around 75 per cent of all plant and animal species then living on Earth. Chicxulub's most famous victims were the dinosaurs – those giant vertebrates that had dominated the biosphere for 150 million years. It wasn't the first time something like this had happened. The dinosaurs themselves came to prominence in the wake of an earlier mass extinction, and there were at least three others prior to that. They weren't all necessarily caused by impact events – there are other possible causes – but it's likely that at least some of them were.

If a 10-km object can cause so much devastation, what about a 1-km one – or even 100 metres? That's not going to wipe out whole species, but it could still be enormously destructive. The object that created Arizona's Meteor crater was only about 50 metres in size. It doesn't need much imagination to visualise what would happen if a similar object hit a densely populated city today. An object of similar size – probably slightly larger – entered the atmosphere over Russia in 1908. This one exploded at high altitude, so there wasn't any crater, but the resulting fireball scorched a huge area of forest, and the blast flattened millions of trees for a radius of 30 km. By a stroke of luck, this happened over the sparsely populated Tunguska river valley, so there were virtually no casualties. If the same thing had happened over Moscow, 3,600 km to the west, it would have been a different story.

In earlier times, before the notion of cosmic impacts was fully understood, an event like Tunguska would probably have been mis-recorded as some other type of natural disaster, such as an earthquake. Did people notice any correlation between such events and 'things seen in the sky'? There has always been a strong association between comets and impending disaster, though more likely this was out of pure superstition.

In many cultures, a comet was once the archetypal 'bad omen'. To give just one famous example, a bright comet appeared in the year 1066, shortly before the Norman invasion of England. One of the scenes in the Bayeux tapestry – created by the Normans after their victory – depicts the English king Harold being warned about it. His fate was sealed; the comet was taken as a certain portent of defeat (conveniently ignoring the fact that the Normans would have seen the comet too).

Halley's comet depicted in the Bayeux tapestry.

In hindsight, we now know that the comet of 1066 was Halley's comet – the largest and most spectacular of the regularly appearing 'short-period comets'. Its orbit brings it back to the inner Solar System time and again, roughly every 76 years. It has been seen since antiquity, but the fact that each appearance was the same object was only worked out after Isaac Newton developed his theory of gravity in the 17th century. Among other things, this explains how objects in the Solar System all move on regular orbits.

Newton was a professor of mathematics at Cambridge University, but it was a professor at Cambridge's great rival, Oxford, who first demonstrated the tremendous predictive power of Newton's theory. This was Edmond Halley, from whom the famous comet takes its name. After observing a comet in 1682 and calculating its orbit, he realised that it was exactly the same comet that had been recorded on at least two previous occasions, in 1531 and 1607. He extrapolated the orbit forwards to work out that the comet would appear again in 1758 – which it duly did, 17 years after Halley's death.

This new understanding hardly made cometary paranoia go away – it just changed its nature. To quote Carl Sagan and Ann Druyan:

> The brand of mischief that comets are said to bring – flood, darkness, fire, rending the Earth asunder – changes with time and astronomical fashion. But the association of comets with catastrophe remains curiously steady through the generations.

Even Halley wasn't immune. In 1694, he presented a paper to the Royal Society called 'Some considerations about the cause of the universal deluge'. Despite wrapping it up in the technical-sounding term 'universal deluge', what he was talking about here was nothing other than the Biblical Flood. In his paper, he ascribes this to 'the casual shock of a comet or other transient body' – using the word 'casual' in its original sense of occurring by chance.

For modern readers, the reaction to any talk about the Bible outside a theological context is likely to be a rolling of the eyes – but Halley was a product of his time. Nevertheless, his description is surprisingly modern, even hinting at two of the currently accepted effects of impacts: tsunamis ('the great agitation such a shock must necessarily occasion in the sea') and cratering ('such a shock may have occasioned that vast depression of the Caspian Sea and other great lakes in the world'). In common with most of his contemporaries, Halley accepted the Bible as an accurate account of historical events. From that perspective, he concludes that the impact hypothesis 'may render a probable account of the strange catastrophe we may be sure has at least once happened to the Earth'.

... and it can happen again

Another follower of Isaac Newton was William Whiston – a literal follower in this case, since he succeeded Newton as professor of mathematics at Cambridge. Like Halley, Whiston was convinced that a comet had caused the Biblical Flood – and he went a step further. He thought the world was due for another disaster of similar proportions. With a comet on its way in 1736, Whiston predicted that it would collide with Earth and destroy civilisation on 16 October that year. As scaremongering exercises go, this one was quite effective. It's said that people fled London for the countryside, banks were so inundated by people wanting to withdraw money they had to close, and in the end the Archbishop of Canterbury was forced to issue a call for calm.

A much more accomplished follower in Newton and Halley's footsteps was the great French physicist Pierre-Simon Laplace. His Wikipedia page lists more than 30 scientific theories and methods under the heading 'known for'. In his book *The System of the World*, first published in 1796, Laplace speculated that cometary impacts might result in global extinctions:

> The greater part of men and animals drowned in a universal deluge, or destroyed by the violence of the shock given to the terrestrial globe; whole species annihilated; all the monuments of human industry reversed; such are the disasters which a shock of a comet would produce.

Although Laplace was taken seriously on most subjects, this proved to be an exception. The scientific consensus in his day, and for almost two centuries afterwards, was that there was no place for sudden catastrophes – caused by comets or anything else – in Earthly affairs. Ironically, this particular dogma originated as a reaction against outdated religious narratives like the Biblical Flood. Having dismissed such things as superstition, science embraced a new paradigm called 'gradualism' – the deliberate polar opposite of catastrophism. As recently as 1972, *The Penguin Dictionary of Geology* boasted the following entry:

> **Catastrophism.** The hypothesis, now more or less completely discarded, that changes in the Earth occur as a result of isolated giant catastrophes of relatively short duration.

Having been tossed aside by mainstream science, catastrophism found a new home in the realm of pseudoscientific cranks and religious doom-mongers. This simply created a vicious circle, further hindering its consideration by serious scientists. The most notorious case was that of Immanuel Velikovsky in the 1950s. A qualified psychologist – but not a qualified astronomer – he ascribed a wide range of historically recorded disasters to cosmic collisions, using a narrative that was almost wilfully ignorant of the way the Solar System actually works. The timescales he talked about were those applicable to human affairs, not astronomical or geological processes. A good (and mercifully brief) summary of his theory was provided by science fiction author and pseudoscience-debunker John Sladek:

> Between 1500 and 700 BC, the Earth was visited by a series of comet-induced catastrophes, which Velikovsky has choreographed thus: Jupiter collides with Saturn, knocking a piece out of itself which becomes a comet. The comet collides with the Earth several times (causing earthquakes, floods, meteor showers, etc). It then collides with Mars, knocking it out of orbit. Mars bears down on us (more quakes etc). Finally, Mars and the comet collide again, very near the Earth. Small comets are pulled off the comet's tail; they become the asteroid belt, while Mars is knocked back into orbit, and the comet settles down to become the planet Venus.

That was such utter nonsense it hardened the science community more strongly than ever against catastrophism. At the same time – and frustratingly for scientists – Velikovsky's ideas proved enormously popular with a certain section of the general public. A whole new sub-genre of pseudoscience grew up around it, on a par with – and catering to the same audience as – flying saucers and alien abductions.

Velikovsky-style catastrophism is still alive and well today, in the form of scaremongering rumours that pop up every now and then on the internet. In 2012, for example, a number of people became convinced that a collision with a non-existent planet called Nibiru was imminent. Such stories, originating in small online communities, can sometimes reach much wider audiences thanks to irresponsible reporting by tabloid newspapers. To take just one example, in January 2017 the *Daily Mail* carried the headline: 'A doomsday asteroid will hit Earth next month and trigger devastating mega-tsunamis, claims conspiracy theorist'.

An important point needs to be made here. The *Daily Mail* didn't run that headline simply for the benefit of other conspiracy theorists – there just aren't enough of them to make it a commercially viable proposition for a large-circulation newspaper. Instead, they ran it for the millions of ordinary people who see the whole thing as an object of humour. It was an entertainment piece, not a scaremongering one. For scientists, this 'giggle factor' – that's the term they use – is just as much of an annoyance as Velikovsky-style crankery. That's because, at some point towards the end of the 20th century, the scientists themselves stopped laughing at catastrophism.

It started with the discovery of the Chicxulub crater, and the unravelling of its cause-and-effect relationship to the demise of the dinosaurs. Just as eye-opening was another impact event, which anyone with a telescope could witness for themselves. This was comet Shoemaker-Levy 9, which crashed into Jupiter in 1994. The impact produced scars in that planet's atmosphere the size of Earth – and it sounded the final death-knell of dogmatic gradualism. Even the most sceptical scientist could picture what would have happened if Shoemaker-Levy 9 had hit Earth instead of Jupiter.

Taking the threat seriously

There's now a conscious, worldwide effort to locate and track near-Earth objects – the collective name for any asteroids or comets that might pose an impact risk in future. As a result, we're much more likely than previous generations to have advance warning of a collision. Whether that makes us any less vulnerable is a different matter altogether. So is there anything we could do to avert a collision if we saw it coming?

In principle, the answer is yes. The movies *Armageddon* and *Deep Impact* are packed with bad science, but their central idea – that a carefully planned space mission could deflect or destroy an incoming space object – is perfectly sound. That gives us a huge advantage over the generations, and indeed the species, that came before us. As Neil deGrasse Tyson puts it:

The dinosaurs didn't have a space programme, so they're not here to talk about this problem. We are, and we have the power to do something about it. I don't want to be the embarrassment of the galaxy, to have had the power to deflect an asteroid, and then not, and end up going extinct.

Much the same sentiment has been expressed by others. The science fiction writer Arthur C. Clarke, for example, put it very succinctly: 'The danger of asteroid or comet impact is one of the best reasons for getting into space.'

But is the equation really as simple as that: space travel equals the end of cosmic impacts? Before we can answer that, there are several other questions we need to look at first: what these hazardous objects are, where they come from, how much damage they can cause – and how to find them in the first place.

ROCKS IN SPACE

2

In pre-scientific times, when opinions on most subjects were governed by religious beliefs, the idea of rocks falling from space was frowned on for the simple reason that rocks are earthly and the sky is heavenly – so there can't possibly be any connection between the two. Although that was the established view of medieval Christian scholars, it didn't originate with them; the ancient Greek philosopher Aristotle was a strong proponent of the same idea. He maintained that all rocks had to originate on Earth, and if they appeared to fall from the sky, it was because they'd been thrown about by a strong wind.

By the start of the 18th century, the world was a much more scientific place. Thanks to Newton, it was known that the planets, moons and comets of the Solar System all obeyed the same physical laws as everyday objects on Earth. Most of the old superstitions about outer space disappeared – except for one. People still refused to believe that rocks fell from the sky.

In 1769, the French chemist Antoine Lavoisier presented a paper to the Académie Royale des Sciences on the subject of 'a stone which it is claimed fell from the sky'. He concluded that it was an ordinary terrestrial rock that had been struck by lightning. Several books quote him as saying, bluntly and dogmatically, that 'there are no stones in the sky, therefore stones cannot fall from the sky'. Sadly this memorable phrase doesn't appear anywhere in his paper, and is probably apocryphal. Nevertheless, it neatly sums up the prevailing attitude in Lavoisier's time.

The prevailing attitude among the scientific cognoscenti, that is. Plenty of ordinary people believed that stones fell from the sky, for the perfectly good reason that they'd seen them fall. Time and again, scientists were forced to shrug off eyewitness reports as the misperceptions of ignorant peasants. When a hail of meteorites fell near the French town of Agen in 1790, sworn affidavits attesting to the fall were provided by 300 witnesses. When these were published in a scientific journal, the editors appended a note saying 'we do not place any faith in any of them'.

By that time, however, a revolution was afoot – the French revolution. Before long it was a guillotinable offence to talk about 'ignorant peasants'. When the physicist Jean-Baptiste Biot collected another set of witness statements in 1803, he gave them all equal weight irrespective of social status. Every name was prefixed by the same honorific: *Citoyen* or 'Citizen'.

This new, unbiased sample was enough to convince Biot of the extraterrestrial origin of the fallen rocks. 'I have succeeded in putting beyond doubt one of the most astonishing phenomena that mankind has ever observed,' he said. That wasn't an idle boast, it was a factual statement backed up by a wealth of data. The science community– which is more susceptible to overwhelming evidence than its detractors think – quickly agreed with him. Today, Biot is generally acknowledged as the person who established the reality of meteorites.

That word 'meteorite' was coined soon after Biot published his findings, as a derivative of the older word 'meteor'. The difference between the two can be confusing for newcomers to astronomy – with other words like comet and asteroid often adding to the confusion. Even for subject-matter experts, the boundaries between the terms can get very blurred. The problem, as with a lot of astronomical jargon, is that the names were coined before people fully understood what they were talking about. Perhaps it's best if we go right back to basics.

Apparitions in the sky

Meteors are a common sight in the night sky, looking like sudden, brief streaks of light. Known since antiquity, they would have been even easier to see in the days before ubiquitous street lighting. For a long time, they were assumed to be purely atmospheric phenomena, and therefore in the domain of meteorology (as the name suggests) rather than astronomy.

In a sense, meteors *are* an atmospheric phenomenon. They are the result of cosmic debris entering the upper atmosphere at high speed, where the material is raised to an extreme temperature like a re-entering spacecraft. Some of the larger chunks reach the ground, in which case the surviving fragments are referred to as meteorites. However, the great majority of meteors are just microscopic specks of dust. People rarely give any thought as to what happens to these, but it's an interesting question even so.

Of course, being hit by a microscopic speck of dust isn't going to make anything go extinct, so we're straying off topic a little here – but it's worth a brief detour. The first point to make is that, even if they're only specks of dust, there are an awful lot of them. In an average day, about 100 tonnes of meteor dust falls on the planet.

So where does it all go? In the first instance, it's spread evenly all over the surface. Anything that falls on the sea is obviously lost right away, and over most of the land it soon leaches into the soil. So one of the best places to find it is on non-porous surfaces like city rooftops and gutters. The upshot is that you don't need to go to a museum to see a meteorite. The sludge in your gutter almost certainly contains a few particles that came from outer space.

Okay, so now we know that most meteors are tiny specks of dust. But where did those specks come from? Some of them are primordial bits of the Solar System that never got swept up into larger objects, like moons and planets (well, the bits in your gutter did get swept up into a planet eventually). Another significant proportion of meteor dust comes from comets. People who have only ever seen comets in photographs sometimes confuse them with meteors, but they're actually a different visual experience altogether.

Like meteors, comets have been seen since ancient times, but they're much rarer. It's usually only a few times per century that a comet is visible to the naked eye – rather than several times a night in the case of meteors. And while meteors come and go in the blink of an eye, a comet hangs in the sky for several days before disappearing. A typical comet looks like a bright star with a fainter smudge of tail – or, with a bit of imagination, like a star trailing a long mane of hair. That's where the name comes from, since *kometes* in Greek means 'long-haired'.

Most of a comet's mass is concentrated in something much too small to be seen with the naked eye: a nucleus of rock and ice typically a few kilometres across. As the comet's orbit brings it close to the Sun, some of the ice and other volatile chemicals evaporate off. It's this process that makes it so visible to the naked eye, creating a roughly spherical coma of dust and gas around the nucleus, and a long tail blown off by the solar wind. That's not really a wind in the terrestrial sense, but a constant stream of fast-moving particles emanating from the Sun – but the effect isn't too different. It's a small part of that blown-off material in the tail that eventually finds its way to Earth in the form of meteor dust.

There's one other type of object that's going to feature prominently in this book. Unlike meteors and comets, these were completely unknown before the invention of the telescope. They were given the name asteroids – Greek for 'star-like' – because that's exactly how they look through a telescope. On the other hand, they behave just like tiny planets, slowly drifting from one day to the next against the backdrop of stars. In fact, 'tiny planet' is an almost perfect description of an asteroid.

Asteroids can be almost any size – it's just a question of defining sensible upper and lower boundaries. Off the top end of the scale is Ceres – close to 1,000 km in diameter, and with a roughly spherical shape like a planet. When it was first discovered it was, in fact, called a planet, but then for many years it was classed as the largest of the asteroids. By current definitions, it's a 'dwarf planet' like Pluto. This hands the 'largest asteroid' title to Vesta, which is just over 500 km across and the archetypal potato shape of most asteroids.

The lower end of the asteroid size scale is even more arbitrary. The current figure is one metre, with anything smaller than that being described as a 'meteoroid'. That's because most meteorites come from objects in this size range, although a few would be classed as small asteroids by current definitions.

The more asteroids, comets and meteors/meteorites/meteoroids are studied, the more their properties are found to overlap. There are comet-like asteroids and asteroid-like comets. In 1949, comet hunters Albert Wilson and Robert Harrington discovered a comet that was duly named after them. Three decades later, asteroid hunter Eleanor Helin discovered asteroid 4015. It turned out to be exactly the same object, comet Wilson-Harrington. But neither 'discovery' was a mistake – 4015 Wilson-Harrington really does tick all the boxes as both a comet and an asteroid.

There's one fundamental thing that asteroids, comets and meteoroids all have in common: they travel in orderly orbits around the Sun. That's what planets do, of course, so while we're running through some basics we ought to take a brief look at those too. Most people will be familiar with the general arrangement of the Solar System, with the Sun at the centre and eight planets moving in roughly circular orbits around it. The least circular – i.e. the most elongated – of these orbits belongs to the innermost planet Mercury, with an eccentricity* of 0.2. Another similarity, besides their near-circularity, is that all eight orbits lie in approximately the same plane. Taking the plane of Earth's orbit – called the ecliptic – as a baseline, the highest inclination (Mercury again) is a mere 7 degrees.

The average distance from the Earth to the Sun is about 150 million km, or (and this is an easier number to remember) precisely 1 astronomical unit (AU). The latter gives us a nice, convenient yardstick for measuring cosmic distances. In these terms, Mercury is 0.4 AU from the Sun, Venus 0.7 AU, Mars 1.5 AU, Jupiter 5 AU, Saturn 10 AU, Uranus 19 AU and Neptune 30 AU.

So where do space rocks fit into it? Sci-fi movies often give the impression they drift around aimlessly, but the fact is that they follow precisely defined orbits just like the planets. The majority of the asteroids live in the asteroid belt between the orbits of Mars and Jupiter, while comets – the regularly appearing ones, at least – spend most of their time in the Kuiper belt beyond the orbit of Neptune.

---

* 'Eccentricity' is a technical term we'll use quite a lot in this book. Its mathematical definition is complicated, but all you really need to know is that it's a number between 0 and 1 which measures how much an orbit departs from circularity, with 0 being perfectly circular and 1 so stretched that the orbiting body escapes.

If that was the end of the story, there'd be nothing to worry about. Comets and asteroids would never come close enough to the Earth to pose a collision risk. Yet we know – from all those craters – that they sometimes do. How do they manage to do that, if they're moving in planet-like orbits? To answer that question, we need to look a little more closely at how orbits work.

Celestial mechanics

The term 'celestial mechanics' was coined by Laplace – it's the title of another of his books – and refers to the application of Newton's theory of gravity to the motion of objects in the Solar System.

There's an oft-repeated anecdote about Laplace's book on celestial mechanics that you may have come across. When Newton first applied his theory of gravitation to the Solar System, the solution he obtained was unstable – so he suggested that God occasionally had to intervene in order to keep it from falling to pieces. Using superior mathematical methods, however, Laplace showed that the Solar System was inherently stable after all. In the famous anecdote, the Emperor Napoleon asks Laplace why his book doesn't mention Newton's God-intervention theory, to which Laplace replies 'I had no need of that hypothesis'. He was simply saying that he didn't need to invoke God to explain the stability of the Solar System, although the remark is often presented as an example of scientific arrogance, with Laplace claiming to have mathematically disproved the existence of God.

What Laplace did, in fact, was to strengthen Newton's argument that it's basically gravity that makes the Solar System go round. Gravity comes from mass, and 99.8 per cent of the Solar System's mass is concentrated at its centre, in the Sun itself. This creates a powerful 'gravity well', which pulls everything else down into it. If an object had zero tangential motion – i.e. movement at right angles to the direction to the Sun – there would be nothing to stop it falling all the way to centre. Of course, because the Solar System is billions of years old, everything that met that criterion fell into the Sun long ago. So everything that's left does have a significant tangential motion, and that's all that's needed to keep an object from falling into the Sun. It's still pulled in by gravity, but the combination of that force with its tangential velocity causes it to swing round the Sun in a closed orbit.

Under these circumstances, the orbit follows a shape called an ellipse. Essentially this is a stretched circle – in fact a circle is just the special case of an ellipse with zero eccentricity. In a circular orbit, the motion is always tangential, with no component of velocity towards or away from the Sun. In the more general case of a non-circular ellipse, that's only true at two points on the orbit: the furthest point from the Sun, called the aphelion, and the closest, the perihelion. We'll try to avoid technical terms wherever possible, but those are two we're going to need over and over again.

To soften the blow of introducing two multisyllabic Greek words, here's an amusing piece of trivia. If you saw the word aphelion on its own, without knowing what it meant, you'd probably run the p and h together and pronounce it 'afelion'. But seeing it in conjunction with perihelion, it's clear that '-helion' is a suffix (meaning Sun) and 'ap-' is a prefix (meaning away from) – so you might pronounce the two parts separately: 'ap-helion'. That's how most astronomers say the word, but it turns out they're wrong. You were right the first time – the correct pronunciation, according to most dictionaries, is afelion.

The idea that a closed orbit in the Solar System always takes the form of an ellipse – as opposed to any other shape – is a direct consequence of Newton's theory of gravity. All Newton did, though, was to provide an explanation for a fact that was already known from observation. In 1609, more than 30 years before Newton was born, the German astronomer Johannes Kepler worked out that the planets all move on elliptical paths – that's the first of his three laws of planetary motion.

That may sound rather vague, but it's a very powerful insight. An ellipse is defined by just five numbers, called orbital elements: the distance to the Sun at perihelion, the eccentricity, the inclination to the ecliptic plane, and the angular directions (the astronomical equivalent of longitude) to the perihelion and the point where the orbit crosses the ecliptic. That means that given a small number of measurements, the entire orbit can be predicted in a precise, mathematical way.

Kepler's second law says that a planet moves faster when it's close to perihelion than when it's further from the Sun. Again, this isn't just a vague assertion – Kepler expressed it quantitatively as 'the line joining the planet to the Sun sweeps out equal areas in equal intervals of time'. As for the third law, that says the time needed to complete an orbit is longer for larger orbits. That may sound pretty obvious, but once again Kepler formulated it in a rigorously mathematical way (readers of a nervous disposition may want to look away at this point): the orbital period increases in proportion to the square root of the cube of the average value of the perihelion and aphelion distances.

That sounds unnecessarily complicated, but it's great for astronomers because it allows them to predict planetary motions with a high degree of accuracy. In fact, Kepler's laws would give an exactly correct answer if there were only one planet in the Solar System. With multiple planets they're still a pretty good approximation, but if you want an exact solution in that case you have to turn to Newton's more general law of gravity.

Kepler's laws are a remarkable achievement, because he worked them out purely by trial and error – testing various mathematical models against observations until he found something that fitted. That's different from the approach used by Newton (and most later scientists), who had an underlying physical theory to guide them. Kepler, in fact, wasted a lot of time trying to explain the Solar System in terms of metaphysical ideals such as Platonic solids (regular shapes that fascinated the philosopher Plato) and the so-called 'music of the spheres' before he hit on his now-famous laws.

Although they're called laws of planetary motion, Kepler's laws apply just as well to anything else that orbits around the Sun, including asteroids and comets – and even Elon Musk's midnight cherry Tesla Roadster (giving a whole new meaning to the phrase celestial mechanics). When

Musk's company SpaceX tested its new Falcon Heavy launch rocket in February 2018, they wanted to prove to potential customers that it was capable of sending a payload beyond Earth orbit, without all the hassle and expense of developing a real payload. So they used a ten-year-old car belonging to the CEO – who, of course, was also CEO of Tesla Motors. The result was arguably the most spectacular publicity stunt the Solar System has ever seen.

The original intention was to send the car out to the orbit of Mars (not the planet itself, which would be at another point in its orbit when the car got there). You might think that involves aiming outwards, since Mars is the next planet out from the Sun. It turns out that it's a lot easier to aim at right angles to the obvious direction: tangential to the Earth's orbit, in the same direction that the planet is already moving around the Sun. It does that at a rate of one revolution every year, which equates to about 30 km/s. By launching in the same direction, the spacecraft – sorry, the car – gets the full benefit of that enormous speed, completely free of charge. Together with the additional boost provided by the Falcon Heavy, the Tesla ended up on an orbit of higher energy than the Earth.

Since it was travelling tangentially to the Sun when it was launched, that point must represent the perihelion of its new orbit. Thanks to its higher energy, however, it can climb further out of the Sun's gravity well to a much larger aphelion. The plan, originally, was that this aphelion would lie somewhere on the orbit of Mars.

To be honest, though, SpaceX didn't worry too much about hitting a particular aphelion. This was just a test, and they simply gave the car as big a boost as possible. As it turned out, this was more than they needed for Mars orbit, and the car will still have plenty of outbound speed when it gets there. Its momentum will keep it going towards the inner fringes of the asteroid belt, as the following diagram shows.

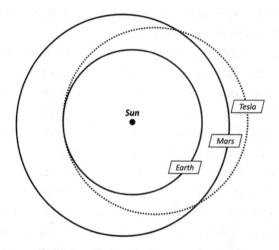

Approximate orbit of Elon Musk's Tesla Roadster, relative to the orbits of Earth and Mars.

All this may seem like a digression from the subject of the book – but it isn't. If launching an object from Earth with an excess of energy can produce an aphelion in the asteroid belt, what about the mirror-image situation? If an object in the asteroid belt loses energy somehow – for example through a gravitational encounter with another asteroid – its perihelion can be reduced to the point that it becomes a collision hazard with Earth. So any attempt to understand asteroid-like orbits is time well spent.

The vermin of the skies

In Kepler's time, only the first six planets were known. Their distances from the Sun, measured in astronomical units, run (as given previously) like this: 0.4, 0.7, 1, 1.5, 5, 10. There's something annoyingly unsatisfactory about that jump from 1.5 (Mars) to 5 (Jupiter). As a ratio of consecutive numbers, it's much bigger than any of the others. That led Kepler to speculate that there must be an undiscovered planet in that gap, to smooth the sequence out. It was an idea other astronomers agreed with. Much later, after the discovery of Uranus at 19 AU seemed to confirm the pattern, a concerted effort was made to find the missing planet.

In 1800, a loose confederation of more than 20 astronomers set to work, calling themselves the 'celestial police'. The laws they were upholding were Kepler's, and unlike most policemen they were hunting for a suspect that obeyed those laws. They found one, too – but not in the way they were expecting. Rather than a single elusive planet, they discovered that the gap between Mars and Jupiter is filled with a large number of sub-planetary objects: the asteroid belt.

First there was Ceres – the largest of all, and now classified as a dwarf planet. The second largest, Vesta, was only the fourth to be discovered, but it's actually the brightest of the asteroids, thanks to its light coloration (it can even be seen with the naked eye, if you know when and where to look). In spite of that, it's still just a star-like point of light when seen from Earth. Even through the Hubble space telescope, Vesta is a tiny, almost featureless blur.

Asteroids are more remarkable for their quantity than their size. Astronomers were constantly discovering new ones throughout the 19th century, even when they weren't looking for them. It got so bad they started to refer to asteroids as 'the vermin of the skies'. In 1912, the American astronomer Joel Hastings Metcalf complained that:

Formerly, the discovery of a new member of the Solar System was applauded as a contribution to knowledge. Lately it has been considered almost a crime.

Today, the total number of asteroids over 1 km in size is estimated at up to two million. That sounds a lot, but remember how small they are. The total mass of the asteroid belt is only about 4 per cent that of the Moon – or less than 3 per cent if you don't count Ceres. The significance of this low mass is that the asteroids don't have enough gravity to pull themselves together into the single planet Kepler predicted to exist between Mars and Jupiter.

The belt extends from roughly 2 AU – about a third of the way beyond Mars – to 3.5 AU, about a third short of Jupiter. Most of the asteroid belt objects move on roughly circular orbits – for example both Ceres and Vesta have eccentricities less than 0.1. Orbital inclinations are typically somewhat higher than planets, with Vesta at 7 degrees to the ecliptic (the same as Mercury), Ceres at 10 degrees, and others as high as 30 degrees.

In terms of composition, like most of the inner Solar System asteroids tend to be rocky. Further out, on the other hand, the Solar System becomes increasingly icy. There's no mystery here – it's colder simply because it's further from the Sun. There's a slight catch though, because astronomers use the word 'ice' in a different way to everyone else. They don't just mean water ice – though that's an important part of it – but any substance that exists in frozen form in the outer Solar System but would melt or vaporise closer to the Sun. That includes methane, ammonia, carbon dioxide and many hydrocarbons. All these chemicals become increasingly common beyond the orbit of Jupiter – and really come into their own in the Kuiper belt that lies beyond Neptune.

The Kuiper belt, named after the Dutch-American astronomer Gerard Kuiper, is big – much bigger than the familiar part of the Solar System inside the orbit of Neptune. The latter extends from 0 to 30 AU, while the Kuiper belt carries on from 30 AU to at least 100 AU. If you hold a blank CD or DVD up to the light, it's mostly opaque but you can see through the middle of it. That's partly because of the hole in the centre, but there's also a transparent plastic region around this called the stacking ring. The outer edge of this ring is slightly raised, and if you let this represent the orbit of Neptune, then all the other planets would be found inside the transparent area. As for the rest of the disc, including the all-important data area – that's the Kuiper belt.

Like the asteroid belt, the Kuiper belt is made up of many small objects, but because of its larger size there are even more of them: billions rather than millions. The Kuiper belt contains a few objects large enough to be seen from Earth through telescopes – including the famous dwarf planet Pluto, which is more than twice the size of Ceres. Most

Kuiper belt objects, however, are very small – of a similar size and shape to asteroids, but more interesting in composition because they contain 'ices' (in the astronomers' sense of the word) as well as rock.

The inner part of the Kuiper belt is relatively stable, with objects moving on orderly, near-circular orbits. Its outer part – called the scattered disc – is significantly less so. This is important to us, because a small gravitational perturbation in this region can knock objects onto deeply plunging orbits which take them all the way down into the inner Solar System. From our point of view, when they arrive, we see them as comets. As they come closer to the Sun, its heat causes volatile components to boil off, forming the characteristic tails that make comets so easily visible.

This picture of the Kuiper belt as the birthplace of comets evolved gradually over the course of the 20th century. The naming of the belt after Kuiper himself, who wrote about it in the 1950s, is fairly random, since he was just one of several people who contributed ideas on the subject. In Kuiper's time, it was all pure speculation – which was only put on a firmer footing when the first Kuiper belt objects (other than Pluto) were discovered in the 1990s.

Like the asteroid belt, the Kuiper belt lies close to the central plane of the Solar System, although it's a somewhat thicker doughnut shape. As a consequence, comets that originated there generally have low orbital inclinations. Their orbits also bring them into the inner Solar System fairly frequently, from which they get the collective name of 'short-period comets'. In the case of comet Encke, the period really is short – just 3.3 years, corresponding to an aphelion that's actually inside the asteroid belt. What distinguishes it as a comet, rather than an asteroid, is its icy composition and its high orbital eccentricity of 0.85.

Most short-period comets are disappointingly unspectacular in appearance. That's because they've been round the Sun so many times most of their volatiles have already boiled off. Encke, for example, is hardly ever visible without a telescope. At the other extreme, the brightest of all the short-period comets is also the most famous: Halley's comet. It travels on an even more elongated orbit, with an eccentricity of 0.97 and aphelion way out in the Kuiper belt at 35 AU. Its orbital inclination is a relatively high 18 degrees – or, if you want to be pedantic, 162 degrees (180 minus 18), since Halley goes round the Sun in the opposite direction to almost everything else. Its orbital period, as mentioned in the previous chapter, is 76 years.

Which begs the question – if 76 years is a short period, what's a long period?

Surprise visitors

Short-period comets like Halley and Encke are as predictable in their movements as planets and asteroids. We've seen them before, so we know when they're going to make their next appearance. Other comets, however, come as a complete surprise – we see them once, and we never see them again. These (as you've probably guessed) are called long-period comets, with aphelion far beyond the Kuiper belt, and an orbital period measured in thousands of years.

There's another difference, too. Because long-period comets are generally newer and fresher – in terms of the number of times they've been round the Sun – they have more volatiles on board to make a spectacular tail. The brightest long-period comets – often referred to as 'great comets', for obvious reasons – completely dominate the night sky, and in some cases can even be seen in daylight. There are usually one or two great comets per century – the most recent was Hale-Bopp in 1997. That particular comet has a period of 2,500 years and an aphelion 370 AU from the Sun (which may not sound much if you say it quickly – but remember that 1 AU is about 150 million km).

As we saw in the last chapter, there's a longstanding superstition that comets are portents of bad things happening on Earth. Sadly, that proved to be all too true in the case of Hale-Bopp, which really was the harbinger of a terrestrial tragedy. In anticipation of the comet's arrival, 39 members of a Californian religious cult called Heaven's Gate committed ritual suicide. This wasn't the usual end-of-the-world comet panic, but something much more bizarre – as David Barrett explains in his book *New Believers*:

When comet Hale-Bopp approached Earth in March 1997, one photograph was thought to show a small dot in its wake; as a hoax, someone suggested that an alien space-craft was following behind the comet. Through so-called 'scientific remote viewing', two separate people claimed to have contacted the aliens on board the spacecraft and reported that they were benevolent. The members of Heaven's Gate made video recordings expressing their joy that the spacecraft had come to collect them, and that they would be able to leave their human shells behind. There is no indication in the videos of worry, fear or sadness that they were about to die.

Of course there was no spaceship, and Hale-Bopp was a perfectly ordinary long-period comet. One of the features of such comets – besides their incredibly elongated orbits – that distinguishes them from more familiar Solar System objects is the fact that their orbital inclinations are completely random. In the case of Hale-Bopp, for example, its orbit was tilted at almost 90 degrees to the plane of the ecliptic.

While short-period comets come from the ring-like Kuiper belt, long-period comets originate in a roughly spherical shell surrounding the Solar System at an even greater distance. It's called the Oort Cloud, and it extends from around 2,000 AU to at least 100,000 AU, almost halfway to the next nearest star. Its existence is hypothetical – there's no way we could observe it directly – but it's believed to be made up of trillions of comets which spend most of their time on near-circular orbits. At that distance, however, the Sun's gravity is very weak, so it's not too difficult for a few comets to be knocked onto more eccentric orbits – for example by the gravitational effect of other stars. They then fall into the inner Solar System where we see them as long-period comets.

Unlike the Kuiper belt, the Oort Cloud is fully deserving of its name. The first clear description of it appeared in a paper by the Dutch astronomer Jan Oort in 1950

– a description which still comes impressively close to our present-day understanding:

> From a score of well-observed original orbits it is shown that the 'new' long-period comets generally come from regions between about 50,000 and 150,000 AU distance. The Sun must be surrounded by a general cloud of comets with a radius of this order, containing about 100,000,000,000 comets of observable size; the total mass of the cloud is estimated to be of the order of 1/10 to 1/100 of that of the Earth. Through the action of the stars fresh comets are continually being carried from this cloud into the vicinity of the Sun.

Any orbit that takes an object from the Oort Cloud – or from the Kuiper belt or the asteroid belt, for that matter – 'into the vicinity of the Sun' is necessarily going to cross the Earth's orbit. And that raises the stakes enormously, because it means we could be on a collision course.

COLLISION COURSE 3

'Sir, the possibility of successfully navigating
an asteroid field is approximately 3,720 to 1'
C-3PO in *The Empire Strikes Back*

In science fiction, colliding with an asteroid is an ever-present
hazard of space travel. Space rocks are portrayed as densely
packed and moving in unpredictable, haphazard ways – neither
of which is true of the real-world asteroid belt. Nevertheless,
the asteroid collision trope is so appealing that even Isaac
Asimov – normally a stickler for scientific accuracy – employed
it in his first published story, 'Marooned off Vesta' (1939). One
of its characters asserts that 'bucking the asteroids is risky business', while another says: 'we should have avoided the asteroid
belt by plotting a course outside the plane of the ecliptic'.

Actually that's not necessary – there's very little risk
involved in crossing the asteroid belt. Yes, there are a lot of
asteroids – millions of them – but they're spread through
an enormous volume of space. A cube of the same volume

would have sides something like 500 million km in length. That gives each asteroid an awful lot of space to itself. As asteroid specialist J.L. Galache puts it:

> Each asteroid gets a cube of space with sides measuring roughly 515,000 km. That will be the average distance between asteroids.

On top of that, the orbits of the larger asteroid belt objects are so well known that any half-competent space-farer could plot a course to avoid them. In the real world, at least eight spacecraft have passed through the asteroid belt – and they've all emerged unscathed (just as well when you think that at the speeds involved, being hit by even a tiny meteoroid-sized object could cause substantial damage).

The key to survival is that it only takes a spacecraft a year or so to pass through the danger zone. That's the blink of an eye in astronomical terms. It would be a different matter altogether if it had to sit there for century after century, just waiting for a collision. That's what we do here on Earth.

Hazardous orbits

For a space object to pose a hazard to the Earth, it has to be travelling on an orbit around the Sun which crosses the Earth's. In numerical terms, that means it must have an aphelion greater than 1 AU and a perihelion less than 1 AU. That's the case for many comets – including Halley, Encke and Hale-Bopp – and thousands of asteroids.

The existence of Earth-crossing asteroids is a relatively recent discovery. The 19th-century astronomers who complained about the 'vermin of the skies' did so because they got in the way of more serious observations, not because they were seen as a physical threat. The situation began to change in 1932, when an asteroid was discovered with an aphelion at 2.3 AU in the asteroid belt, but a perihelion of just 0.65 AU, well inside the orbit of the Earth. It was given the name Apollo, and other objects on similar orbits are called 'Apollo asteroids' (a designation that can be a little confusing, since there's no connection at all to the Moon-landing Project Apollo – but the asteroid family got first dibs on the name by several decades).

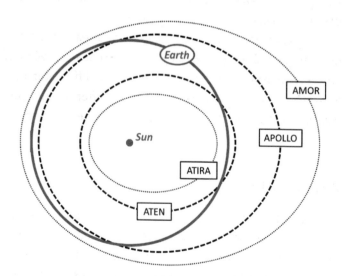

Types of near-Earth asteroid. Thick dashed lines show the two Earth-crossing types, while thin dotted lines are non-Earth-crossing types.

If you cast your mind back to the previous chapter, you may recall that Kepler's second law tells us that objects move faster at the perihelion end of their orbits than at aphelion. That means that asteroids like Apollo spend most of their time outside Earth's orbit, where they move at a relatively leisurely pace, and then whiz along much faster for the brief time they're inside it (the same is true of comets). We can use another of Kepler's laws here, too – the third, which tells us that larger orbits take more time to complete a revolution than smaller ones. This gives us another characteristic of Apollo asteroids: they take more than a year to complete an orbit around the Sun.

There's also another class of Earth-crossing asteroids that go round the Sun on smaller orbits than the Earth, allowing them to complete the circuit in less than a year. The reason they still cross the Earth's orbit – often by just a small fraction of an astronomical unit – is because their paths are more eccentric than the Earth's. The first asteroid of this type – with perihelion at 0.8 AU and aphelion at 1.1 AU – was discovered by Eleanor Helin in 1976, as part of a systematic search for near-Earth objects. It's called Aten, and it gave its name to this general class of asteroids.

You may remember Eleanor Helin's name from the previous chapter. She was the one who rediscovered comet Wilson-Harrington and realised it also qualified as an asteroid. Aten and Wilson-Harrington are only the tip of the iceberg, though. Helin was one of the most successful of all asteroid-hunters, with over 900 discoveries to her credit. As well as Aten itself, she also discovered the second asteroid in the Aten class, in September 1978. That was the same month as the Camp David peace agreement between Israel and Egypt, and in recognition of that Helin gave the newly discovered asteroid the unusual name of Ra-Shalom. Ra was the ancient Egyptian sun god, while Shalom is a Hebrew word meaning peace.

As asteroid names go, Ra-Shalom is one of the most imaginative. At the other end of the scale, it's a shame for those of us who like to abbreviate names to their first letter that Aten (and the class named after it) begins with the same letter as Apollo – and there's worse to come. There are two other types of near-Earth asteroids, which don't cross the Earth's orbit but come close enough to it to be considered potential collision risks in future – and these groups begin with an A too.

Amor-type asteroids always lie outside Earth orbit, but their closest point – i.e. perihelion – is less than 0.3 AU away. Amor itself was the first such object discovered, in 1932 – the same year as Apollo. Its perihelion is 1.1 AU, while its aphelion is at 2.8 AU in the asteroid belt.

For a long time, that was all there was – Apollos, Atens and Amors. Logically, however, there ought to be another type of near-Earth asteroid, which spends all its time inside Earth orbit. The problem with such objects is that they're intrinsically difficult to see, because they tend to lie close to the Sun in the sky. It wasn't until 2003 that the first such object was discovered, with perihelion at 0.5 AU and aphelion just inside Earth orbit at 0.98 AU. Following the established trend, they gave it a name starting with A – Atira – and that's also the name of this class of asteroids.

Of course, simply having two objects travelling on orbits

that cross each other isn't enough to make a collision. Both objects have to be in exactly the same place at exactly the same time. As with the spacecraft crossing the asteroid belt, the sheer scale of astronomical distances comes to the rescue here. In the course of a year, the Earth travels about 940 million kilometres – getting on for 75,000 times its own diameter – at a speed of 30 km/s. That means there are a lot more ways that an asteroid can miss Earth than hit it, even if their orbits do intersect. Even a close, non-collision, encounter is a rarity – much to the disappointment of astronomers, who like to see such things.

Of course, there's one close encounter everyone wants to see – and that's Halley's comet. It comes round roughly once in every human lifetime – but whether that's a close or distant pass depends on the position of Earth in its orbit when the comet arrives. Now, you might think that working out where the Earth is in its orbit is a difficult technical problem, suitable only for a fully qualified celestial mechanic – but actually it's nothing more complicated than the calendar date.

It turns out that the orbit of Halley's comet crosses the Earth's at the May and October positions. That means you get a good view of it if it happens to arrive close to one of those months, and a less good view at other times. You can get an idea of this from the diagram opposite, which shows the position of Earth when the comet reached perihelion on its last five visits. It was pretty close when Halley himself saw it in September 1682, and again in November 1835 and April 1910. However, it was further away in March 1759, and really quite distant for the only appearance anyone reading this book might remember, in February 1986.

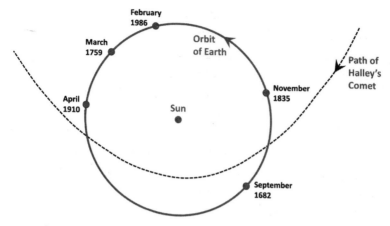

Position of Earth at the perihelion of Halley's comet.

There's another factor that make collisions less likely than they might appear at first sight. The plane of the Earth's orbit, called the ecliptic plane, is less than a degree away from the central plane of the Solar System – but most comets and asteroids move on significantly more tilted orbits. Halley's comet, for example, with its inclination of 18 degrees, doesn't pose a serious collision threat even when it crosses the Earth's orbit. It's well above the plane of our orbit on its incoming trajectory and well below it on the outgoing leg.

For a significant number of near-Earth objects, it's possible to define a clear-cut 'minimum orbit intersection distance', or MOID. The object is prevented from coming any closer than this by basic geometrical considerations. For Halley's comet the MOID is 10 million km, for Apollo it's 4 million km, and for Aten it's 17 million km. For comparison, the distance from the Earth to the Moon is around 0.4 million km, so those are all pretty safe miss distances.

Does that mean we can delete all the objects with a non-zero MOID from our list of collision hazards? Sadly, the answer is no – because orbits can change.

Moving the goalposts

Physics has a guilty secret most popular science books don't let you in on. They'll tell you that Newton showed how all orbits are governed by his law of gravity – but in itself that isn't as useful as it sounds. It's a statement about theory – and attempting to put that theory into practice leads to some of the most intractable problems in physics. In the jargon of mathematics, you have to 'integrate' Newton's equations before you can make any practical use of them.* That's not a problem if all you have is a single planet orbiting a single star. That's a case for which Newton's equations can be integrated exactly – and the answer is Kepler's three laws of motion (which we knew already, as it happens).

The problems start to appear when we go on to the next step, adding a third body to the first two. That might not sound like a big step – especially if the third body is much smaller than the first two – but it's still a step too far. The resulting equations just can't be integrated – not in any general way. Newton couldn't do it, Laplace couldn't do it, and we still can't do it today.

---

* Integration is a mathematical operation which, for example, can calculate the distance an object travels in a given time based on a knowledge of its velocity.

The only option is to work through the equations on a case by case basis – and you never quite know what's going to happen. Many people will have heard of chaos theory: the idea that complex systems, such as weather patterns on Earth, are highly sensitive to the tiniest of changes. But the fact is, you don't even need a complex system to see that sort of thing. As long ago as the 1880s, the French mathematician Henri Poincaré found similar behaviour in certain solutions to the gravitational three-body problem. To make matters worse, some of the orbits he came up with weren't even periodic; seeing an object on one pass said nothing about where it would be on the next pass. As Ian Stewart wrote in his book *Does God Play Dice?*:

> In human affairs, two's company and three's a divorce. In the same way, in celestial mechanics the interaction of two bodies is well behaved, but that of three is fraught with disaster.

If the three bodies maintain fairly constant distances from each other – as in the classic case of the Earth, Moon and Sun – the situation isn't too bad, with only minor perturbations from Keplerian orbits. But the more eccentric one of the orbits gets, the more it complicates the situation. In the case of an Apollo asteroid, for example, with a perihelion in the neighbourhood of the Earth but an aphelion out towards Jupiter, the gravitational pull of the latter planet can cause gradual changes in the asteroid's orbit. It's exactly this effect that threatens to cause disaster in the novel *Shiva Descending*, written in 1979 by astrophysicist Gregory Benford in collaboration with William Rotsler. 'Shiva' is a fictional Apollo asteroid that is paid no attention because its originally high orbital inclination posed no possible threat to Earth. Over the years, however, it's been quietly dipping down until its MOID is zero – and Shiva is on a collision course with Earth.

With comets, the situation is even worse. As they fall in from the Kuiper belt or beyond, their highly elongated orbits can take them very close to Jupiter or one of the other massive outer planets. There's nothing gradual about the effect in this case. The encounter changes the comet's orbit immediately, in a way that's exactly analogous to the 'gravitational slingshot' effect used by some spacecraft. More formally known as a gravity assist manoeuvre, this can be used to change the direction of a spacecraft – or even transfer energy to it in order to accelerate it to higher speed. The technique was originally proposed in the 1960s by a mathematics student named Michael Minovitch, and later put to practical use by NASA in its Pioneer and Voyager probes of the 1970s. NASA's PR people, being Americans, manage to explain the whole thing by means of a baseball analogy:

> Picture a fast pitch coming toward the batter. The baseball represents a spacecraft. Now picture what the batter does, swinging a bat with all the force he or she can muster. The business end of the bat in motion represents a massive planet like Jupiter. The bat connects with the ball: wham! The ball receives momentum from the bat, and takes off in a different direction with a lot more speed as it soars out of the stadium.

Under the right circumstances, exactly the same thing can happen with comets. In this case, the effect is unplanned – and it's just as likely to decrease a comet's energy as increase it. That's a bad thing from our point of view, because it pushes the comet further into the inner Solar System. Something like that happened in the case of Hale-Bopp, which passed very close to Jupiter in 1997. 'Close' in this context means about 100 million km – around four times the distance between Jupiter and its outermost satellite. That encounter had a significant effect on Hale-Bopp's orbit, reducing its aphelion from over 500 AU to about 370. While its previous appearance was all of 4,200 years ago, its next will be a mere 2,500 years in the future.

Okay, that's no immediate cause for alarm – but Jupiter encounters can affect short-period comets too. After Halley's comet, probably the most famous of these is 67P/Churyumov–Gerasimenko – the one that was visited in 2014 by the Rosetta spacecraft and its Philae lander. Two centuries ago, no one knew anything about 67P, because it never came anywhere near the Earth. Its perihelion was around 4 AU from the Sun, just inside the orbit of Jupiter. Since that time, however, a series of gravitational encounters with that planet have progressively reduced the perihelion. It's now down to 1.2 AU – making it officially a near-Earth object (the defining boundary for which is 1.3 AU).

For a spacecraft, use of a gravity assist manoeuvre is a bonus option. The usual way to change from one orbit to another is by means of rocket thrusters. These use another of Newton's laws: the one that says action and reaction are equal and opposite. Ejecting mass in one direction, in the form of rocket exhaust, produces a force on the spacecraft in the opposite direction. This can be used to speed it up, slow it down or change its course, depending on the direction of the thrust. Asteroids can't do that, because they're just lumps of rock. But what about comets? Their most distinctive feature is the way they eject matter, in the form of dust and evaporating volatiles, as they approach the inner Solar System. That necessarily produces a force on the comet, just like a small rocket motor.

The most notorious 'rocket-powered' comet is the one discovered by American astronomers Lewis Swift and Horace Tuttle in 1862. During that same appearance, observers in South Africa noticed a strange sideways drift in its orbit, which was at odds with Newtonian gravity. Soon afterwards, French astronomers spotted material spurting out of the side of the comet, exactly correlated with the anomalous changes in its motion. As Gerrit Verschuur wrote in his 1996 book *Impact!* – 'In the jargon of the space age, those little bursts produced midcourse corrections'.

This was of more than academic interest, because comet Swift-Tuttle was following a potentially dangerous trajectory, with perihelion just inside Earth's orbit. Its aphelion was way out in the Kuiper belt at 50 AU, giving it an even longer period than Halley's comet – about 130 years. That meant it wouldn't be seen again until shortly before its next perihelion passage in 1992. Initial observations on that occasion produced an orbit calculation of the 'good news/bad news' variety. The good news: Swift-Tuttle wasn't going to hit the Earth on this occasion. The bad news: there was a small but non-zero chance it *would* hit the next time around, in 2126.

Fortunately, the comet's built-in rocket power came to the rescue. As it came closer to the Sun over the coming months, new jets appeared – just like those seen in 1862 – and its orbit changed again, so there was no longer any possibility of a collision in 2126. As the *New York Times* put it: 'Doomsday has been postponed a thousand years or so, at least as far as comet Swift-Tuttle is concerned.'

Artificial Space Hazards

The Space Age has brought us a brand new hazard from the skies, in the form of falling space hardware. In the absence of atmospheric drag, an object in Earth orbit would remain there indefinitely. At the relatively low altitude where most satellites operate, however, there is still enough of an atmosphere to slow them down and eventually bring them back to Earth. While small satellites will burn up on re-entry, there's a real hazard from larger objects. The usual procedure with these is to perform a controlled de-orbit manoeuvre and bring them down in a specially designated 'spacecraft cemetery' – a point in the Pacific Ocean that is more than 2,000 km from the nearest land. The largest object that's been disposed of in this way was the Russian space station Mir, which was deliberately de-orbited in March 2001. With a length of 30 metres and a mass of 130 tonnes, it would have posed a serious risk if it had crashed to Earth in an uncontrolled way.

Not all spacecraft re-entries go according to plan. The worst accident occurred in January 1978, when Russia was still part of the Soviet Union. One of its spy satellites,

Cosmos 954, malfunctioned and crashed to Earth exactly where it wasn't supposed to, in north-west Canada. At less than four tonnes it wasn't in the same league as Mir, but it posed a major hazard for a completely different reason – it was powered by a nuclear reactor containing 50 kg of fissile uranium. The crash spread radioactive debris over thousands of square kilometres, making it one of the most hazardous space impacts in history.

Cometary debris

If comets eject material as they swing past the Sun, where does it all go? The ejected particles obey Kepler's laws just as much as the comet itself – so the answer is they follow orbits similar to that of the comet. They're not quite the same, because the particles are ejected in various directions at various speeds. It may help at this point to visualise how a comet tail works.

As you can see from the graphic opposite, there are actually two separate tails, with dust particles and gas behaving differently. The dust tail is the more visible of the two, because it reflects sunlight. That's not true of the gas, but a fraction of it is ionised – like the gas inside a neon sign – which causes it to glow with its own light.

You might expect a comet's tails to trail behind it as it moves through space, but actually they always point away from the Sun – even on the outbound leg, when that involves pointing in the direction of travel. As Carl Sagan and Ann Druyan explain:

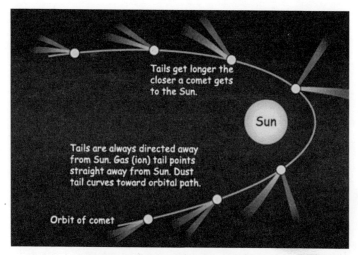

Tails get longer the closer a comet gets to the Sun.

Sun

Tails are always directed away from Sun. Gas (ion) tail points straight away from Sun. Dust tail curves toward orbital path.

Orbit of comet

Schematic illustration of a comet tail.
NASA image

The tails of comets are much more like the effluvia from an industrial smokestack blown back on a windy day than the long hair of a bicyclist flowing behind her as she coasts down a hill on a windless day; it is not the motion of the comet through some resisting gas that determines the orientation of the tails, but rather something like a wind blowing out of the Sun.

That 'wind blowing out of the Sun' consists of a stream of charged particles – mainly protons and electrons – which constantly flows out from the Sun at high speed. It really is a high speed, too – typically around 400 km/s. For comparison, the speed of a satellite in orbit around the Earth is a mere 8 km/s, and the Earth itself travels round the Sun at 30 km/s.

The existence of the solar wind was only established in the 1950s. Until then, the fact that comet tails always point away from the Sun had been something of a mystery to astronomers. An earlier suggestion was that the tails are pushed away from the Sun by radiation pressure – a phenomenon predicted by the Scottish physicist James Clerk Maxwell in the 19th century. Sunlight has a small but perceptible momentum of its own, and therefore exerts a tiny pressure on anything it comes into contact with. In fact radiation pressure does play a role in the behaviour of comet tails, but a less important one than the solar wind.

As it moves away from the comet, the dust tail forms a gradual curve, since the particles that make it up are all travelling in their own personal orbits at slightly different speeds. On the other hand, the ionised gas, being much lighter, is swept along more by the solar wind than gravity – so the gas tail remains straight.

On at least one occasion, the existence of a gas tail caused yet another brand of comet-related panic. When Halley's comet appeared in 1910, although the comet itself came no closer than 20 million kilometres, it was predicted that Earth would pass right through its tail. In itself, that might have passed without much fuss – but spectroscopic observations made at the same time showed the presence of a chemical called cyanogen in the tail. As the name suggests, that's closely related to the deadly poison cyanide. With a little encouragement from the mass media, the world's population was gripped by panic. Churches held prayer vigils, while opportunistic entrepreneurs cashed in selling anti-comet pills.

Strangely, it wasn't just the tabloid press who promoted the erroneous notion that there was some sort of danger to the public. The French astronomer and science populariser Camille Flammarion got in on the act too. As the *New York Times* reported at the time: 'Professor Flammarion is of the opinion that the cyanogen gas would impregnate the atmosphere and possibly snuff out all life on the planet' – to which it added: 'most astronomers do not agree with Flammarion'. In reality, the cometary gases were much too diffuse to have any physiological effect, even if the Earth had passed through the densest part of the tail (which it didn't). Anyone affected would breathe in only a few molecules of cyanogen, as opposed to the sextillions needed for a lethal dose.

As for the dust tail, the fact that its particles all have slightly different orbital velocities – with some moving a bit faster than the comet and others (the ones ejected backwards) a bit slower – means that over the centuries the dust gradually spreads out to fill the whole orbit. Because the particles have sideways motions as well as fore-and-aft, they end up occupying a thick tubular volume in space. In some cases, this tube can intersect the Earth's orbit – usually at two distinct points per comet.

Whenever the Earth passes one of these points – which it will do as regularly as clockwork in its annual journey around the Sun – the result is a meteor shower, with a sudden surge in the number of comet-dust meteors lasting a few days. Referring back to the orbit of Halley's comet shown on page 41, it's clear that the associated meteor showers ought to occur in early May and late October – which indeed they do. The first shower is called the Eta Aquariids, the second the Orionids.* If a grain of meteor dust happens to fall on your roof at either of those times, it's probably a piece of Halley's comet.

There's another meteor stream associated with comet Encke, which produces meteor showers a short time after Halley's – the Beta Taurids in June and the Taurids around Halloween. This particular stream contains an unusual amount of debris, because Encke – with its short orbital period – has been around the Sun so many times. In fact it's probably the relic of a much larger comet, which broke up at some point in the past. While Encke is the largest of the remaining fragments, the same stream also contains several asteroids (which are members of the Apollo group, because they cross the Earth's orbit and have a period greater than a year).

The fact that there's so much material in the Encke-Taurid stream – with some of it in big chunks – makes it a significant hazard to our planet. It's likely, for example, that the object that exploded over Tunguska in June 1908 was part of the Beta Taurid shower. We briefly touched on that incident in the opening chapter, but we're far from finished with it. Now that we've set the scene, it's time to get down to business. What actually happens when an asteroid or comet hits the Earth?

---

* Meteor showers always appear to radiate from a specific location in the sky, determined by the relative motion of the Earth and the meteor stream – and this is how they get their names. In the case of the Eta Aquariids the radiant is near the star Eta Aquarii, while for the Orionids it is in the constellation of Orion.

DEATH FROM THE SKIES 4

The 1977 novel *Lucifer's Hammer*, by Larry Niven and Jerry Pournelle, describes the worldwide devastation that would be caused by a cometary impact on Earth. Yet prior to the impact, many of the characters remain unconcerned about the threat. As a TV interviewer says at one point, 'You told me that a comet, even the head, is largely foamy ice with rocks in it. And even the ice is frozen gases. That doesn't sound dangerous.' At first sight, this scepticism seems valid enough. How can a chunk of rock and ice, even if it's a kilometre across, pose a serious threat to a planet that's over 12,000 km in diameter? The key lies in a formula that most people learn in school: the kinetic energy of an object is equal to half its mass multiplied by the square of its velocity. That tells you the energy the comet is carrying purely by virtue of its motion through space – and all of that energy is suddenly transferred to the Earth at the moment of impact. Even a relatively small mass becomes very dangerous when it's multiplied by the square of a big number.

There's another formula that physicists use, which tells them the speed an object needs in order to stay in a circular orbit at a given distance from the Sun. In the case of the Earth, the formula gives – as near as makes no difference – 30 km/s. Yet another formula – a more important one in the present context – yields a number that is the square root of two (approximately 1.4) times larger than this, or 42 km/s. This is the speed an object would have as it passes Earth, after being dropped towards the Sun from the very edge of the Solar System. In other words, it's a reasonable estimate of the speed of a long-period comet falling in from the Oort Cloud. Anything travelling slower than 42 km/s would have to originate closer to the Sun, while anything faster wouldn't be gravitationally bound to the Solar System.*

Combining Earth's orbital speed of 30 km/s and a typical comet speed of 42 km/s gives a relative speed of 72 km/s in the worst case of a head-on collision. In practice, it's more likely to be an oblique encounter, so it wouldn't be quite that bad. As one of the characters in *Lucifer's Hammer* puts it: 'Depends on the geometry of the strike. Shall we say 50 km/s as a reasonable closing velocity?'

When you square 50 km/s and multiply it by half the comet's mass – maybe a billion tonnes – you get an awful lot of kinetic energy. We could work it out in joules (the scientist's preferred unit of energy) or kilowatt-hours (a more familiar unit to most people), but the answer would be so big it wouldn't really convey anything. Instead, the most

---

* We're going to need that number 42 again in the next chapter. Readers familiar with *The Hitchhiker's Guide to the Galaxy* should have no difficulty remembering it.

common unit used to describe impact energies is the megaton (MT) – the energy that would be released if you exploded a million tons of TNT.

The term originated during the Cold War, when the majority of nuclear weapons had explosive yields of the order of a few megatons. The largest ever exploded – the Soviet Tsar bomb of 1961 – managed 50 MT, but that was a carefully controlled test, so it didn't really give much idea of the destruction such a bomb could cause. The nuclear bombs everyone knows about – the ones that flattened the cities of Hiroshima and Nagasaki at the end of World War Two – were really quite small by Cold War standards, at around 0.02 MT each. If you add all the non-nuclear bombs that were dropped in the course of the same war, you get up to a total of about 3 MT.

These human-made explosions are puny in comparison to what nature can do. A good benchmark in the latter context is the volcanic eruption that came close to destroying the island of Krakatoa in 1883. That was equivalent to about 200 MT – big enough to have global consequences as well as local ones. The eruption blasted dust and gas into the upper atmosphere, lowering average temperatures around the world for the next several years.

Now that we have a destructiveness scale we understand, where does a cosmic impact sit on it? Let's start small, with an Apollo asteroid a hundred metres in diameter – and there are something like 100,000 of those – hitting the Earth at a closing speed of 25 km/s. That equates to a kinetic energy – half mass times velocity squared – of around 100 MT. That's two Tsar bombs, half a Krakatoa ... or 5,000 Hiroshimas.

As for larger impacts, let's go back to that fictional comet in *Lucifer's Hammer*. At just over a kilometre across, and travelling at an estimated 50 km/s, its kinetic energy comes out at an eye-watering 600,000 MT – or, as one of the characters in the novel puts it, 'about 3,000 Krakatoas'.

History lesson

Impacts were much more common in the early Solar System, when there were a lot more space rocks around. The best place to see the effects of these early impacts is the Moon. The features that are most easily visible from Earth – the dark patches known as 'seas' – are actually huge impact scars that were filled in by lava billions of years ago.* Even the most recent of them was probably formed before life had appeared on Earth almost four billion years ago. After that, however, impact events continued to occur on a smaller scale – and the very phrase 'lunar landscape' is synonymous with impact craters.

One of the Moon's craters is prominent enough to be seen with the naked eye from Earth. This is Tycho – about 85 km across, and surrounded by rays of bright material extending over a much larger area. At one time all the lunar craters would have had such rays – caused by material ejected during the impact itself – but they gradually fade away over time. Tycho's rays are still visible because it's a relatively 'young' crater – only about 100 million years old.

---

* The Moon has cooled over the course of billions of years and is no longer volcanically active, but it was when the 'seas' were formed.

The full Moon photographed with an ordinary DSLR camera, clearly showing the crater Tycho near the southern edge.
Author's photograph

Looking at the Moon from Earth – even through a telescope – it's easy to think its landscape is static and unchanging. If a crater that is 100 million years old can be described as 'young', does that mean impacts are a thing of the past, and they simply don't happen any more? Not too long ago, astronomers might have thought that – but now we know that new Moon craters are appearing all the time. NASA's Lunar Reconnaissance Orbiter started its task of mapping the Moon in 2009. Since that time, it's seen at least 200 new craters appear – all of them 10 metres or more in diameter – which simply weren't there the first time it looked. On one occasion in March 2013, NASA scientists even captured the flash of an impact, and correlated it to a newly formed crater.

That may or may not have been the first direct observation of a lunar impact. More than 800 years earlier, in June 1178, the chronicler Gervase of Canterbury wrote:

> After sunset when the Moon had first become visible a marvellous phenomenon was witnessed by some five or more men who were sitting there facing the Moon. Now there was a bright new Moon, and as usual in that phase its horns were tilted toward the east; and suddenly the upper horn split in two. From the midpoint of this division a flaming torch sprang up, spewing out, over a considerable distance, fire, hot coals, and sparks.

In 1975, the planetary scientist Jack Hartung wrote a paper suggesting that what the group had actually witnessed was a meteorite impact on the Moon. He even linked it to a specific crater, called Giordano Bruno, which is 20 km in diameter and in a location which fits Gervase's description. Not everyone in the science community was convinced by Hartung's explanation, but it remains a possibility – and an intriguing one.

Compared to the Moon, impact craters on Earth are a rarity. Part of the reason is that small objects – the kind that might produce one of those 10-metre craters on the Moon – break up in the Earth's atmosphere before they hit the ground. As for larger objects, many of them fall in the sea, which covers more than 70 per cent of the planet. The fraction that hit land do make craters, but the craters don't last long due to various processes that happen on the Earth but not on the Moon. Some of these are obvious – such as atmospheric weathering or getting covered in vegetation – while others are more subtle. The Earth, unlike the Moon, has a large hot core – due in part to the greater abundance of radioactive elements – and this drives a continuous cycle of geological activity which tends to smooth out the surface of the planet.

As a result, the Earth only has a single impact crater that really looks like one. As already mentioned, it's in Arizona, and to make up for the fact that there's only one of them, it has two names. Sometimes it's called Barringer crater, sometimes Meteor crater. It's just over a kilometre across, and almost 200 metres deep. It was formed about 49,000 years ago, by an iron meteorite about 50 metres across (that ratio of 20 to 1 between crater diameter and impactor size is fairly typical, so it's worth remembering). The explosion in that case would have been Tsar-bomb sized, around 50 MT.

For a long time, the impact crater in Arizona was the only one anyone knew about. With no real interest in the subject of impacts among the scientific community prior to the last decade of the 20th century, no one really bothered to look. Once they did start looking, however, they found a lot more – and around 200 are known today. Some of them are very large – a hundred kilometres or more across – but most of the large ones are very old, with ages measured in millions of years rather than thousands.

The Barringer, or Meteor, crater in Arizona.
US Geological Survey

Among the very largest of Earth's impact craters is Chicxulub, centred just off the coast of the Yucatan peninsula in Mexico and spreading for 200 km – partly across land, partly out to sea. Applying the '20 to 1' rule of thumb, this suggests the crater was caused by an object about 10 km across. The Chicxulub impact has been dated to 66 million years ago – which is significant, because it coincides with the extinction of the dinosaurs.

On the geological timescale, that particular extinction corresponds precisely to the end of the Cretaceous period. That may sound like a spooky coincidence, but it isn't really. Geological periods are defined according to the fossil species found in them, and a lot of species happened to disappear all at once at the end of the Cretaceous. The dinosaurs – which weren't a single species but hundreds – were simply the best-known victims. They'd dominated both the Cretaceous and the Jurassic that preceded it – a total span of 135 million years. They were just the tip of the iceberg, though. In all, the close of the Cretaceous period saw the extinction of something like 75 per cent of all plant and animal species on the planet.

Since a large part of the Chicxulub crater is underwater, it might never have been discovered, if it weren't for the fact that the Gulf of Mexico is of great interest to oil companies. That proved to be a two-edged sword, though. On the one hand, it meant that geologists were paid handsomely to make sure the area was surveyed as thoroughly as anywhere on Earth. On the other, the geologists in question were sworn to secrecy – because the oil business is a very competitive one.

As a result, the crater ended up being discovered twice independently – on both occasions by American geologists working under contract for oil companies. The first discoverer was Robert Baltosser in 1966, the second Glen Penfield in 1978. Commercial secrecy wasn't the only problem, either. When Penfield hinted at the crater's existence at a conference of academic geologists, no one showed any interest. In those days, the geological community just didn't want to hear about giant impact craters.

It was different in other branches of science. In the wake of the Apollo Moon landings, space scientists developed a fascination with the subject of lunar cratering. They also worked out that, on the scale of the Solar System, Earth is right next door to the Moon – and a bigger target to boot. If an impact on the Moon 100 million years ago had created the 85-km crater Tycho, what was going on here on Earth at that time? The answer is that it was the middle of the Cretaceous period: any dinosaurs looking up at the Moon would have seen the impact happen.

One of the first people to put two and two together was Harold Urey, a physicist who had worked on the Manhattan Project to design the first nuclear bomb. In the journal *Nature* in 1973, he wrote: 'It does seem possible and even probable that a comet collision with the Earth destroyed the dinosaurs'.

Another member of the Manhattan Project team was Luis Alvarez, who went on to win the Nobel Prize in 1968 for his work on the physics of elementary particles. His multifaceted career also encompassed some of the earliest research on radar, which brought him into contact with a young RAF technician named Arthur C. Clarke. Much later, the latter commented that: 'Luis seems to have been there at most of the high points of modern physics – and responsible for many of them'. The highest of all those high points came in the 1980s, when Alvarez was instrumental in convincing the world that the dinosaurs had been wiped out by an asteroid.

His interest in the Cretaceous extinction came about through his geologist son Walter Alvarez. Between them, they discovered a very odd thing about the boundary between the Cretaceous and the period that followed it (called the Tertiary at the time of their work, but since renamed the Palaeogene). The result was a paper, co-written with chemists Frank Asaro and Helen Michel, that was published in the journal *Science* in 1980. Its straight-to-the-point title was 'Extraterrestrial cause for the Cretaceous–Tertiary extinction'.

What Alvarez and his team had found was that all over

the world, the last Cretaceous rocks and the first Palaeogene rocks are separated by a thin layer containing a much higher concentration of the element iridium than is usual near the Earth's surface. Iridium is one of the densest naturally occurring elements – denser than gold, platinum or uranium – and as a result the Earth's own iridium sank down into the planet's core long ago. Any iridium in near-surface rocks must have come from meteorites. There's nothing outlandish about that – it's something scientists had accepted for a long time. They had, however, visualised a gradual rain of small meteorites over a long period of time, resulting in a uniform distribution of iridium.

What the Alvarez team found instead was a huge spike in iridium deposition at the end of the Cretaceous. The size of the spike even allowed them to estimate the size of the object that had caused it – between 10 and 15 km, a perfect match for the Chicxulub crater.

Even if we assume a relatively modest speed of 25 km/s, the Chicxulub impact would have equated to something like a hundred million megatons. That's five billion times larger than the Hiroshima explosion (at which, as it happens, Luis Alvarez was the official scientific observer). That's an almighty bang – but is it really big enough to wipe out every single dinosaur on the planet?

Dust clouds and tsunamis

Just like a gigantic nuclear bomb, the immediate effect of the impact would be to convert all that energy into heat, light and a tremendous blast wave. Since the origin of the energy is kinetic rather than nuclear, you might think that at least there's no radiation hazard in this case – but there is. The temperatures and pressures created by the explosion would be high enough to produce X-rays, which aren't too different from nuclear gamma rays in their effect on living things.

The devastation would be so much larger than anything in human experience that it's difficult to imagine. If the impact occurred in the sea, the water would boil. If it was on land, vast areas would be ravaged by firestorms. In either case, the seismic shock would produce earthquakes, and the blast wave would produce hurricane-force winds. And that's just the start. To quote Lisa Randall, in her book *Dark Matter and the Dinosaurs*:

> In fact, essentially most of the great movie disaster scenarios (with the exception of a zombie apocalypse) follow in the wake of a sufficiently big impact. The impact itself creates shock blasts, fires, earthquakes and tsunamis. Dust can block out the atmosphere, temporarily ending photosynthesis and eliminating the majority of the food sources for most animals.

The dust thrown up by a Chicxulub-sized impact would prevent sunlight reaching Earth's surface for months – maybe even years – to come. That would be enough to kill off all the plants, disrupt the food chain for most animals, and cause a sudden drop in temperatures over the whole planet. As well as dust, the atmosphere would be filled with pollutants like nitrous oxide and sulphur, causing acid rain to fall over the whole planet. And this time – unlike the spurious Halley scare of 1910 – there really would be poison in the air. There might even be cyanide, in much deadlier concentrations than the tail-passing event caused, but the most likely poisons would be heavy metals such as lead. A toxic atmosphere, deep-freeze temperatures, a sunless night that went on for years – it's not surprising there were extinctions.

Even a much smaller object than the ten-kilometre Chicxulub impactor would cause serious devastation. There aren't many Earth-crossing space rocks that big – but there are thousands that are a few hundred metres across. The effect of one of those hitting the planet could still prove much worse than the most destructive natural disaster, such as an earthquake or volcano.

The eruption of Krakatoa in 1883 has already been mentioned. It was one of the biggest in recorded history, so it's worth looking at its effects in a little more detail. Although it occurred on a small Indonesian island, about midway between Java and Sumatra, the explosion was heard thousands of kilometres away in Australia. It completely transformed the geography of the island – which had an area of 10 square kilometres before the eruption, and less than 4 square kilometres after it – and killed 36,417 people according to the official death toll. The ensuing seismic shock caused the whole planet to reverberate for five days. Dust, ash and sulphur dioxide were blasted into the sky, which turned such a bright red all over the world that fire engines were called out as far away as London.

The size of the Krakatoa explosion has been estimated at 200 MT, comparable to the kinetic energy of a 150-metre asteroid hitting the Earth at a speed of 20 km/s. But kinetic energy has an alarming tendency to go through the roof when you tweak the parameters a bit, because it's proportional to speed squared and size cubed.* That means that simply doubling the size to 300 metres and the speed to 40 m/s – still perfectly reasonable figures for an Earth-crossing asteroid – raises the stakes by a factor of 32 to 6,400 MT. The effects, both in terms of immediate destruction and long-term climate change, go up in proportion.

The fact that over 70 per cent of the Earth's surface is covered by water means that, other things being equal, a water impact is more likely than a land impact. That's good news for us land-dwellers, isn't it? Well, no – it isn't. A random impact in the ocean is going to kill far more people than a random impact on land, because most of the Earth's land surface is sparsely populated – but a lot of people live near coasts. It's time to talk about tsunamis.

The Scottish astronomer Bill Napier – who's something of a specialist in cosmic impacts – also happens to be the author of some pretty awesome technothrillers. The first of these, *Nemesis* (1998), deals directly with his specialist subject. It starts with an intelligence rumour to the effect that an asteroid has been diverted onto a collision course with Earth – and targeted at the United States. If the rumour is true, what's the worst-case scenario? Unless the asteroid could be aimed with pinpoint accuracy – which is a physical impossibility – a strike against mainland America would probably

---

* Think of a sphere of constant density. Its kinetic energy is proportional to its mass, which is proportional to its volume, which is proportional to the cube of its diameter.

end up hitting a relatively unpopulated area. On the other hand, if it was aimed at the 'big, easy targets' of the Atlantic or Pacific Ocean, it would bring death and destruction to an entire seaboard of the United States. As one of the novel's characters explains:

> So half a minute after impact we have a ring of water three or four hundred metres high. Wave amplitude falls as it moves out but you're still looking at a 15-metre wave a thousand kilometres from the impact site. ... An earthquake in Chile in 1960 created ocean waves which travelled over 10,000 miles and killed a lot of people in Japan.

The earthquake referred to here was the biggest in recorded history, with a magnitude between 9.4 and 9.6. Its epicentre was inland, about 35 km from the coast, but its most destructive effect was on the other side of the Pacific – not just in Japan but also in Hawaii, New Zealand and Australia. The cause was a huge surge of water, technically called a tsunami but sometimes colloquially referred to as a tidal wave. Its effect, when it makes landfall, looks just like a sudden, very high tide – although actually it's a wave of ultra-long wavelength.

The most devastating tsunami in recent memory occurred in the Indian Ocean in December 2004. The earthquake that created it was slightly smaller than the one in 1960, with a magnitude between 9.1 and 9.3. It was even more destructive than its predecessor, though, because it was centred on the sea floor close to the heavily populated coasts of Indonesia and the surrounding countries. The death toll was over a quarter of a million, making it the biggest natural disaster of this century and the deadliest earthquake since the 16th century. Hundreds of thousands of buildings were destroyed, and over a million people made homeless. It wasn't just humans who suffered, either – it also resulted in irreparable damage to coral reefs, forests and animal populations.

As it travels across the open ocean, a tsunami wave may not look that big. It's only when it enters the shallow coastal zone that it suddenly grows to enormous size, because there's now much less water to carry the same amount of energy. As a rule of thumb, the wave can be up to 40 times higher when it hits land than it was in deep water. The 2004 tsunami, for example, is believed to have grown from about 0.6 metres to 24 metres.

How does that compare with what we might expect from a cosmic impact? In his 1996 book *Impact!*, Gerrit Verschuur relates impactor size to the likely scale of the resulting tsunami. Using his figures, it turns out that the 2004 tsunami would be duplicated if a 50-metre asteroid crashed into the sea a thousand kilometres from the nearest coast. For the much bigger impact considered by Napier in his novel (he uses a figure of 10,000 MT), the '40 times' rule of thumb give a final tsunami height of 600 metres – half again as high as the Empire State Building. If an asteroid like that really did crash into the Atlantic, it would be goodbye New York – and hundreds of other East Coast cities.

There's a worrying corollary to Napier's idea of a sneak attack by asteroid. A genuine asteroid impact – even a relatively small one – has the potential to be misinterpreted as a nuclear missile strike. If that happened at a time of tension, such as existed during the Cold War, it might provoke a retaliatory attack against the supposed aggressors. The ensuing nuclear exchange could end up multiplying the devastation many times over.

An ongoing threat

In December 1908, the British periodical *Pearson's Magazine* carried a picture of St Paul's cathedral in ruins, accompanied by the following caption:

> If a large comet approached within measurable distance of the Earth, the doom of our world would be sealed. Such tremendous heat would be engendered that everything would spring into spontaneous combustion. The hardest rocks would become molten, and no living thing would remain upon the Earth's surface. Buildings and human beings would be scorched to cinders in seconds.

Of course, that's journalistic sensationalism at its worst – but there's an irony in the date of the piece. That very year had already seen the biggest cosmic impact in recorded history – on 30 June 1908 – and it went completely unreported in the English-speaking media. This was the event mentioned earlier, which occurred in the sparsely populated Tunguska river valley in Siberia, 3,600 km east of Moscow.

The impactor on that occasion may have been a small rocky asteroid or possibly a tiny comet. As mentioned in the previous chapter, it's likely to have been part of the Beta

Taurid meteor stream, containing fragments of comet Encke (or the larger comet it was originally derived from). The result, when the Tunguska object entered Earth's atmosphere, was an explosion of the order of 10–15 MT – typical of a Cold War nuclear weapon. The object's size has been variously estimated at between 30 and 70 metres – the same order as the one that produced the kilometre-wide Barringer crater. That object, however, was made of iron, while this one was rocky. The effect is very different.

It's now known that rocky objects less than 100 metres or so in size are likely to explode due to the stresses caused by atmospheric heating. That means they don't produce any obvious crater or large chunk of meteorite on the ground. Instead, the effects are similar to a type of nuclear explosion called an airburst – a tactic designed to produce ground-level destruction over as wide an area as possible. The idea is that the shockwave from such an explosion, when it reaches the ground, is far more destructive than it would have been if detonation had occurred at ground level. Both the Hiroshima and Nagasaki bombs were airbursts – and the Tunguska explosion was a thousand times more powerful than either of them.

Its most obvious effect was to scorch and flatten trees – 80 million of them – out to a radius of 30 km around Tunguska's 'ground zero'. It's all too easy to imagine the devastation if the same thing had happened over a city. Fortunately, on this occasion, the nearest witnesses were in the village of Nizhne-Karelinsk more than 70 km away. They saw a blinding flash, and then several minutes later – when it finally reached them – heard a rumbling explosion. Some of them were knocked off their feet by the shock wave, which also broke countless windows. There were ground tremors, too, like a small earthquake.

Still, no one was killed, and there was no damage that couldn't be repaired fairly easily, so the villagers didn't make a big deal of it. It was only years later that scientists realised it might have been a massive meteorite impact. An expedition belatedly set off in 1928 to look for a crater, or maybe a meteorite – and failed to find either. What they found instead, 20 years after the impact, was a vast expanse of scorched and flattened forest.

At the time, the scientists couldn't imagine what had happened. Now, however, we do know. A 15 MT airburst creates a searing fireball – which is what set the trees alight – followed by a tremendous blast wave, which is what knocked them over. It also put the fire out, which is why the trees are merely scorched, and not reduced to ash. To describe this blast as a 'wind' is a hopeless understatement. As Gerrit Verschuur points out:

> It takes a very strong wind to blow down trees. Even major hurricanes do little more than uproot some trees. ... The point is that to flatten trees so that they fall neatly point- ing away from the blast centre, the winds would have to be so strong that if a similar event were to happen over an urban area in the United States, say, every house would be collapsed in an instant with frightful loss of life. Estimates of the casualties that would result from a Tunguska-like event in a populated area ... suggest as many as 5 million dead.

Few scientists today would dispute that the Tunguska explosion was caused by impact event. This is a relatively recent consensus, though, stemming from the new understanding of such events that's been developed over the last few decades. For a long time, the lack of a crater at Tunguska was assumed to disprove the impact theory, and for much of the 20th century it was a regular entry on lists of the world's 'great unsolved mysteries'. It gave rise to dozens of wild speculations – and not just from cranks.

In 1973, the prestigious scientific journal *Nature* printed an article by two professional astronomers, A.A. Jackson and M.P. Ryan, entitled 'Was the Tunguska event due to a black hole?' In it, the authors argued that the impactor was not a rocky object but a microscopically small black hole – of a kind Stephen Hawking had predicted a few years earlier. Such an impact, they suggested, would 'create an atmospheric shock wave with enough force to level hundreds of square kilometres of Siberian forest', and yet 'no major crater of meteoric residue would result'.

Just to set the record straight, it's also worth pointing out that – despite the suggestion in many 'unsolved mystery' books – no scientist has ever seriously suggested that Tunguska was the result of a UFO crash. This misconception can be traced to a poorly researched book written in the 1960s by French authors Louis Pauwels and Jacques Bergier, which was translated into English as *The Morning of the Magicians*. Here is the offending passage:

> It is suggested in the reports of the Moscow Academy of Sciences on the explosion of 30 June 1908, that this may

have been caused by the disintegration of an interstellar spaceship.

It's true that the idea's originator was a Russian, but he had nothing whatsoever to do with the Moscow Academy of Sciences. His name was Alexander Kazantsev, and he was a science fiction writer by profession. The notion of a spaceship disintegrating over Tunguska comes from his short story 'Explosion', written in 1946.

In a way, it's not surprising that so many wild ideas were put forward to explain Tunguska. People were theorising in a virtual vacuum, with so little known about impacts, and so few eyewitnesses. A century later the situation was completely different. On 15 February 2013, everyone in the world knew that a spectacular meteor had exploded over the Russian city of Chelyabinsk.

The blazing fireball – as bright as the Sun to some observers – was captured on dozens of mobile phones, dashcams and security cameras, and soon found its way onto the internet. Within 24 hours, the footage had gained millions of views. Eyewitness accounts, numbered in thousands this time, not just a small handful, echoed those from Tunguska. People saw the fireball, they heard the explosion and they felt the shockwaves – equivalent to a minor earthquake of magnitude 2.7. There were even reports of a burning or sulphurous smell.

The actual explosion occurred more than 50 km from Chelyabinsk city, so the sound and shock waves took a measurable amount of time – about 2 minutes – to reach the majority of witnesses. On the other hand, the initial flash, travelling at the speed of light, was virtually instantaneous. It brought people to their windows – and they were still standing there when the blast hit. The result was a lot of broken glass, and thousands of injuries – fortunately most of them minor.

As with Tunguska, the Chelyabinsk meteor exploded in the upper atmosphere at an estimated 20–30 km altitude. It was a smaller object – probably 15–20 metres across – but still the largest since Tunguska, resulting in a blast of about half a megaton. For a nuclear weapon of that size, the optimum air-burst altitude (the height needed to maximise destruction on the ground) is about 1.7 km. At well over ten times that altitude, the Chelyabinsk explosion was pleasingly sub-optimal.

Another good thing about Chelyabinsk was the very low angle at which it entered the atmosphere. Before it did so, its speed was about 30 km/s, comparable to the Earth's speed in its orbit around the Sun. Fortunately, at least a third of that speed was lost through atmospheric drag before the object finally exploded – significantly reducing its megatonnage.

Back in 1928, when a team of scientists went to Tunguska in search of a meteorite, they were looking for a single, very big rock – and they found nothing. At Chelyabinsk, knowing that the object had exploded, people looked for multiple small rocks instead – and they found them. In all, about 50 meteorite fragments were retrieved, totalling around a tonne. By far the largest chunk – weighing around 650 kg and some 60 centimetres across – had buried itself in the mud at the bottom of a lake.

As well understood as the Chelyabinsk event was, it wasn't completely immune from the wacky speculations that had characterised its predecessor. One local resident set up a Church of the Chelyabinsk Meteorite, claiming that the object retrieved from the lake contained an important message that only his psychic priests were able to interpret. And on the day of the meteor itself, the BBC quoted the out-spoken Russian nationalist politician Vladimir Zhirinovsky as saying: 'Those are not meteors, it is Americans testing their new weapon'.

That same article mentions another oddity – and a real one this time. On the same day as the Chelyabinsk event, it was known that another, larger asteroid would pass close to the Earth. Called Duende, its orbit was known with high accuracy. Scientists were certain it would miss the Earth, although it would come closer than the geosynchronous communications satellites that orbit at an altitude of 35,800 km. When the Chelyabinsk meteor turned up earlier the same day, it was inevitable that some people jumped to the conclusion that it was Duende and the scientists had got it wrong. But in reality it was on a different orbit altogether, as is shown in the diagram overleaf.

Duende belongs to the Aten family of asteroids, with most of its orbit lying inside the Earth's, but its aphelion slightly outside. Earth just happened to be in the same place (though fortunately not *exactly* the same place) when Duende crossed its orbit. Chelyabinsk, on the other hand, was an Apollo-type asteroid, on an orbit of higher energy and higher eccentricity. Its aphelion was in the asteroid belt, its perihelion well inside Earth's orbit. In this case, the Earth really was in just the wrong place when it happened to cross our path.

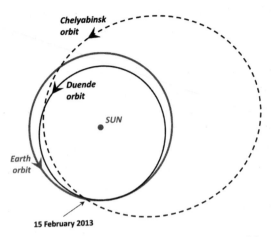

The orbits of two near-Earth objects, one making a close pass on 15 February 2013 and the other entering the atmosphere on the same day.

It was our good fortune that Chelyabinsk was a relatively small object, that Duende missed, and that Tunguska hit in such a sparsely populated area. But is that really such a good thing? According to Gerrit Verschuur, 'civilisation may have suffered the worst piece of luck in its history' when the Tunguska impact failed to wipe out a city. That's a perverse way of putting it, but you can see what he's saying. With no disaster in human history that can unambiguously be ascribed to a cosmic impact,* many people refuse to take the threat seriously. Even so, the last decade of the 20th century

---

* The key word here is 'unambiguously'. There are a number of historic disasters that *might* have been caused by impacts, but there are always other explanations – often more convincing ones.

saw a truly cataclysmic impact event, which would certainly have destroyed civilisation if it had happened on Earth. Instead, it happened on Jupiter.

Comet Shoemaker-Levy 9 was discovered in March 1993 by Carolyn Shoemaker, with assistance from her husband Eugene and from David Levy. As the designation suggests, this was the ninth short-period comet this particular team had found, making it sound run-of-the-mill. But comet number 9 was odd in several ways. It had been captured by Jupiter's gravitational field, so it was orbiting around that planet instead of the Sun. Another effect of Jupiter's gravity had been to break the comet up into a string of more than 20 fragments. And a third oddity came to light when the comet's orbit was worked out, as Gerrit Verschuur explains:

> Orbit calculations showed that in July 1994 Shoemaker-Levy 9 would try to pass within 50,000 km of Jupiter's centre but, because its radius is 71,900 km, this meant the comet was going to try to fly through the planet. That would produce what in the trade is known as a collision.

News of the upcoming collision caused a frenzy of anticipation in the astronomical community. Nothing like this had ever been predicted before. Because the comet was actually a string of objects, it meant multiple impacts – and multiple observing opportunities – over the course of several days. More telescopes were pointed at Jupiter than had ever been trained on the same object in history – all the way down from the Hubble space telescope to amateurs in their backyards.

Even at a distance of 700 million km, the impacts were spectacular – completely belying the 'small objects can't do any real damage' argument. To quote Carl Sagan and Ann Druyan:

> Each of the cometary fragments of Shoemaker-Levy 9 is estimated to have been a few hundred metres to perhaps a kilometre across, plummeting into Jupiter at 60 km/s. A remarkable fact – predicted by almost no one – is that every impact left a dark blemish about the size of the Earth. Each blemish took weeks or months to fade. That a high-speed comet, no bigger than a football field, could produce an effect the size of the entire planet Earth is something worth pondering.

The message is obvious. If Shoemaker-Levy 9 had crashed on Earth instead of Jupiter, the effects would have been global. Yet we knew the collision was going to happen months in advance. If the same thing were true of a terrestrial impact, is there anything we could do about it? That's a question we'll get onto before very long – but first, let's take a step back and look at the big picture.

COSMIC CONNECTIONS 5

'Astronomers should leave to astrologers the task of
seeking the cause of earthly events in the stars.'
The New York Times, 2 April 1985

When the impact theory of dinosaur extinction was first put forward in the 1980s, it was met with scepticism from some quarters. On the one hand, there were the old-school geologists who still bridled at any hint of catastrophism. On the other, there were plenty of non-scientists ready to scoff at the notion that terrestrial events might have extraterrestrial causes. The above quote from a *New York Times* editorial is an example of the latter attitude. It prompted a pithy response from Walter Alvarez and a colleague, which the newspaper printed later the same month: 'It might be best if editors left to scientists the task of adjudicating scientific questions'.

The fact is, though, that many scientists were opposed to the Alvarez hypothesis for exactly the same reason as the *New York Times*. For centuries, people had been brushing off any suggestion of a connection between the Earth and the wider universe as superstitious nonsense. It was the main reason the scientific establishment had taken so long to accept the reality of meteorites. The argument isn't entirely without merit, of course. Astrology – the belief in meaningful correlations between planetary alignments and human affairs – really is nonsense. So was Immanuel Velikovsky's theory, mentioned in Chapter 1, which attempted to explain Biblical events in terms of imaginary collisions between the planets Venus, Mars and Jupiter.

Velikovsky seems to have had a cosy, almost medieval view of the universe, in which astronomical objects are much closer together than they really are. We now know that, to quote *The Hitchhiker's Guide to the Galaxy*, 'Space is big. Really big.' The Juno spacecraft took five years to reach Jupiter – by an admittedly circuitous route – travelling at an average speed of 18 km/s. At that speed, it would take 70,000 years to reach the nearest star beyond the Sun. Given that perspective, it's not surprising that some people have persuaded themselves that the Earth is completely isolated from the rest of the universe.

It's not true, though. There are real connections – hard scientific ones, not woolly astrological ones. Asteroids and comets really do crash into the Earth, and – according to the best current theory – new comets really are knocked in from the Oort Cloud by forces that originate outside the Solar System.

The results aren't always negative, either. As Carl Sagan and Ann Druyan said (paraphrasing the Bible): 'Where the comets giveth, they also taketh away'. Yes, comet impacts can cause mass extinctions – but they're also an essential source of new materials, organic as well as inorganic. It turns out that impacts are an inextricable part of the long-term cycle of life on Earth.

It came from outer space

The idea that the Earth is replenished from space isn't new. Isaac Newton speculated on exactly this subject in the 17th century:

> The vapours which arise from the Sun, the fixed stars and the tails of the comets may meet at last with, and fall into, the atmospheres of the planets by their gravity, and there be condensed and turned into water and humid spirits; and from thence, by a slow heat, pass gradually into the form of salts and sulphurs and tinctures and mud and clay and sand and stone and coral and other terrestrial substances.

The details aren't quite right, but the general principle is valid. As well as 'vapours' (by which Newton meant gas) there's also the dust from comet tails – and, of course, impacts. As Gerrit Verschuur wrote in 1996: 'A dozen or so massive comets carry enough water and organic molecules to provide all the Earth's water and biomass.' While that's true in principle, it's no longer believed that the Earth's entire supply of those things was delivered by impacts – but a significant fraction of it was.

With regard to that word 'organic' that Verschuur used, it can generate a lot of excitement because of its association with living things. To a scientist, however, the word encompasses a whole class of carbon-containing chemical compounds, including the basic building blocks of life, such as amino acids. These are an important constituent of DNA – but saying that a comet contains amino acids doesn't mean it contains DNA.

Organic molecules aren't restricted to comets; they're also found in some asteroids, and in asteroid-derived meteorites. While the great majority of meteorites are dominated by silicon-based chemistry (that's a fancy way of saying they're rocks), about one in twenty contain carbon-based organic compounds as well. They're called carbonaceous chondrites, and from a terrestrial perspective they're much more interesting than their rocky cousins.

One of the most famous carbonaceous chondrites, the 100-kg Murchison meteorite, fell to Earth in 1969 near the village of Murchison in Australia. It broke into several pieces, all of which have been subjected to careful study. The fragments contain at least 15 different types of amino acid, as well as a significant amount of water – about 10 per cent by weight.

We saw in the previous chapter that impacts – and very large ones in particular – were much more numerous in the early history of the Earth. So it's not unreasonable to imagine that they brought much of the raw material needed for the formation and development of life. That's not the same as suggesting that impacts actually brought life itself, fully formed – but it hasn't stopped people speculating that it might have done.

No sooner had scientists accepted the reality of meteorites, than some of them began to wonder if they had been the source of life on Earth. The idea was even given a fancy-sounding name: panspermia.* Here's what Lord Kelvin said on the subject in an address to the British Association in Edinburgh in 1871:

> We must regard it as probable in the highest degree that there are countless seed-bearing meteoric stones moving about through space. If at the present instance no life existed upon this Earth, one such stone falling upon it might, by what we blindly call natural causes, lead to its becoming covered with vegetation.

A few years earlier, in 1864, a carbonaceous chondrite fell near the town of Orgueil in the south of France. Panspermia was a hot topic at the time, and the new meteorite stirred up great interest, with samples being sent to various research institutions. The material was found to be rich in organic compounds, but there was no evidence whatsoever of actual life-forms – or was there?

Fast forward a century, and a team of American scientists rediscovered a fragment of the Orgueil meteorite that had been sealed up in a glass jar long ago, and never previously examined. Astonishingly, it was found to have plant seeds embedded in it. If they had been near the surface, the seeds might have been terrestrial contamination – but they were

---

* Actually the term was coined around 2,300 years earlier by the philosopher Anaxagoras, but he was talking about a different theory.

deep inside, and seemingly protected from the outside world by a glassy fusion layer created by the heat of the meteorite's passage through the atmosphere. Whatever they were, the seeds weren't the result of accidental contamination.

Sadly, the seeds weren't extraterrestrial either. The researchers identified them as belonging to a species of grass indigenous to southern France – and that was just too much of a coincidence. As for the glassy fusion layer – the part of it that covered the seeds was, on closer inspection, found to be dried glue. The whole thing was a hoax – presumably perpetrated by 19th-century researchers against one of their colleagues. The only thing was, the intended victim hadn't bothered to look at this particular fragment.

The idea of panspermia was given a new twist in 1977, when an article appeared in *New Scientist* with the eyecatching title 'Does epidemic disease come from space?'. Its authors, the astrophysicists Fred Hoyle and Chandra Wickramasinghe, argued that the observed patterns of certain diseases are more consistent with an origin in space than with conventional person-to-person contagion. They ascribed the source of these diseases to cometary dust, which they envisaged to contain fully developed viruses. Dormant in the freezing cold of space, these became active after they entered the Earth's atmosphere. Not surprisingly, the theory was met with scorn from both the astronomical and epidemiological communities – but Hoyle and Wickramasinghe were undeterred, and went on to produce a number of popular books on the subject.

Bizarre as it was, the association of terrestrial diseases with comets wasn't completely new. The author of Robinson

Crusoe, Daniel Defoe, also wrote a book called *A Journal of the Plague Year* – a semi-factual, semi-fictional account of the Great Plague that ravaged London in 1665 (when Defoe himself was only five). At one point he notes that 'a blazing star or comet appeared for several months before the plague'. He goes on to speculate on a cause-and-effect relationship – albeit of a purely supernatural kind, since this was long before people understood that diseases were caused by micro-organisms.

The idea there really might be micro-organisms in space was given a short-lived boost by the discovery of tiny fossil-like structures in a meteorite called ALH84001. It was found in the Allan Hills in Antarctica in 1984, but it was only in 1996 that it shot to worldwide fame – with a high-profile announcement by NASA to the effect that a scanning electron microscope had revealed structures resembling terrestrial nanobacteria. The ones in the meteorite were much smaller, however, and there were alternative, non-biological explanations for them – so few scientists were convinced.

Another remarkable – and much less controversial – fact about ALH84001 is that it originally came from Mars. It was hurled into space by an impact on that planet, which gave this particular chunk of Martian rock enough speed to go into orbit around the Sun. That orbit happened to intersect the Earth's, and at some stage the inevitable collision occurred – and the rock came down in Antarctica as a meteorite.

That's not to say the whole sequence all happened in a short space of time. The original impact on Mars would have

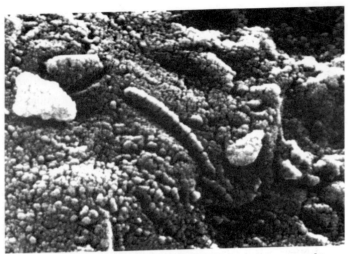

Electron microscope image of the ALH84001 meteorite, showing structures that look like micro-organisms.

NASA image

been millions of years ago, after which the rock had a long career as an Apollo asteroid before it finally crashed to Earth – and even that was more than 10,000 years ago. ALH84001 isn't a particular rarity, either. Meteorites that started life on Mars – or the Moon – are much more common than people used to think. To quote Neil deGrasse Tyson on the subject:

> We conclude that about a thousand tons of Martian rocks rain down on Earth each year. Perhaps the same amount reaches Earth from the Moon. In retrospect, we didn't have to go to the Moon to retrieve Moon rocks. Plenty come to us, although they were not of our choosing and we didn't yet know it during the Apollo programme.

The Alien Octopus

Hoyle and Wickramasinghe's theory about viruses coming from outer space opened the floodgates to a whole new field of speculation. What about more complex organisms – octopuses, for example?

In the Victorian sci-fi classic *War of the Worlds*, H.G. Wells had one character describe the Martian invaders as 'octopuses'. It was a good choice, because octopuses really do look alien to human eyes. They're unusual in other ways, too, with higher intelligence and more complex behaviour than any other invertebrate. When the octopus genome was sequenced in 2015, it was found to possess more protein-coding genes than our own species. Somewhat over-excitedly, the official press release described the result as 'the first sequenced genome from an alien'.

It was meant figuratively, of course – but some people took the idea to heart. In 2018, an academic paper by a number of authors – including Chandra Wickramasinghe – picked up on the fact that so much of that highly complex genome was 'new', in the sense that it wasn't found in any of the octopus's supposed ancestral species. The authors suggested the situation was easier to explain in terms of an extraterrestrial origin – via DNA brought to Earth by cometary impacts – than through the natural process of evolution:

Given that the complex sets of new genes in the octopus may have not come solely from horizontal gene transfers or simple random mutations of existing genes or by simple duplicative expansions, it is then logical to surmise, given our current knowledge of the biology of comets and their debris, the new genes and their viral drivers most likely came from space.

As intriguing as the suggestion is, very few members of the scientific community – other than the paper's own authors – would go along with it.

An even bigger picture

The Moon and Mars are just the start of it. To understand impacts fully, we need to look much further – beyond the Solar System, in fact. This is where we really hit the 'space is big' problem. To help visualise the scale of things, a diagram might help at this point – so let's give it a try.

Everyone knows the Earth is insignificant on a cosmic scale, so we won't even try to include it in our picture. The smallest thing we'll show is the distance from the Earth to the Sun – one astronomical unit. We'll represent it by a dot one millimetre in diameter. On this scale, the aphelion of Pluto's orbit – in the Kuiper belt, where the short-period comets come from – is about 50 mm away. So far, so good – there's no problem fitting that on the page.

We also need to show the Oort Cloud – the source of long-period comets. Its inner edge lies somewhere around 2,000 AU from the Sun. On our scale, that's 2,000 mm – or

2 metres. Things are starting to look bad for our diagram, but it gets worse. The Oort Cloud is postulated to spread at least 50 times further out. That's 100 metres – the length of a football pitch. The nearest star, Proxima Centauri, is further away still – something like 265,000 AU, or 265 metres.

So it looks like we're going to have to manage without that diagram.

As enormous as interstellar distances are, we cannot simply dismiss as irrelevant everything outside our Solar System. In October 2017, an asteroid-hunting team in Hawaii discovered an object close enough to be classified as a near-Earth object – only about 0.2 AU away from us at the time. It soon became apparent, however, that it wasn't going to be 'near Earth' for very long. Its speed as it whizzed past us was around 50 km/s.

Now here's a memory test. Do you remember something about the number 42 from the last chapter? We said that 42 km/s was the maximum possible speed of a comet falling in from the Oort Cloud. There's no room for movement on that number – it's a direct consequence of the conservation of energy, which is something you just can't negotiate with. This new object, whatever it was, had too much energy even for an Oort Cloud object. It didn't belong to the Solar System at all – it came from much further away.

Because the object was first discovered by a telescope in Hawaii, it was given a Hawaiian name: 'Oumuamua, meaning 'advance scout'. There wasn't much time to study it, because it was moving so quickly. It hurtled in from the direction of the constellation Lyra, then whizzed off – after the Sun's gravity had bent its path through 66 degrees

– in the direction of Pegasus. At the time of its appearance, 'Oumuamua was assumed to be an asteroid rather than a comet, because there was no obvious evidence of volatiles. However, subsequent analysis of its motion indicates that it probably was a comet after all, though not a very active one. Whichever it was – asteroid or comet – it was an unusually long and thin one, with an estimated length of 230 metres but a width of only 35 metres.

Almost all the space objects we know are rounder than that – the naturally occurring ones, that is. In the course of their long history, planets and their larger moons get pulled into a near-spherical shape by the strength of their own gravity. With smaller objects like asteroids and comets, the forces holding them together are much weaker – many of them are better described as 'rubble piles' than solid rocks. This means they have a tendency to break up – either due to collisions with other objects or the stresses produced by their own rotation – into potato-shaped objects with an axis ratio of 2 to 1 or less. 'Oumuamua, by contrast, was more like 6.5 to 1.

Those proportions are much more typical of a space vehicle, and there was a flurry of speculation online that that's precisely what 'Oumuamua was – not an asteroid at all but an alien spaceship. Even the scientific community took the possibility seriously enough to look into it. But there were no detectable radio transmissions from 'Oumuamua, and its heat signature was exactly what would be expected from an inert rock. If it was a spacecraft, it was a dead one.

So why was 'Oumuamua so much more elongated than the Solar System's asteroids? One possibility is that it had an unusually high tensile strength – but aside from the spaceship theory, there's no obvious reason why that should be the case. It may simply be that, travelling on its own through interstellar space, 'Oumuamua was never subjected to the kind of collisions that caused ordinary asteroids to break up. Another possibility is that it's actually two normal-shaped asteroids orbiting around each other, giving the illusion of a single elongated object.

As exciting as an interstellar spacecraft would be, that doesn't mean an interstellar rock isn't significant in its own right. The fact that this one passed so close to Earth – and with a huge kinetic energy compared to a Solar System object of the same size – just adds to the list of space hazards we face.

Even without the worry of interstellar intruders, we can't ignore what goes on beyond the Solar System. Remember that when Jan Oort first theorised about the cloud that was later named after him, he remarked that 'through the action of the stars fresh comets are continually being carried from this cloud into the vicinity of the Sun'.

That phrase 'through the action of the stars' might sound like astrological mumbo-jumbo, but Oort had in mind something more tangible. So how could other stars have an effect inside the Solar System? In his paper, Oort talks about the possibility of a passing star distorting cometary orbits through its gravitational effect. That may indeed be part of the answer, but it probably doesn't happen often enough to be the only factor.

To understand why, we need to take another step back and look at the whole galaxy. The name comes from the Greek word for 'milk', in reference to the Milky Way – the blur of light that can sometimes be seen stretching across the night sky. It's a system of about a hundred billion stars, of which the Sun is just one. It may sound like a crowded place – and in a sense it is, but it's also very orderly. Most of the bright stars – including the Sun and its nearest neighbours – are confined to a thin disc, in which they move in roughly circular orbits.

It's a little like a hugely scaled-up version of the Solar System, with one important difference. In the Solar System, most of the mass is concentrated in the Sun, right at the centre, plus a few big planets like Jupiter and Saturn. In the galaxy, on the other hand, the mass is spread out pretty evenly between all those hundred billion stars. As a result, they tend to move in a smoothed-out gravitational field, without really noticing each other as individuals. Close encounters of the kind needed to perturb the Oort Cloud do occur, but not very often. There is, however, another possibility.

First, we need to get a better idea of the scale of the galaxy. To avoid using ridiculously large numbers, we'll introduce another astronomical unit of measurement – the light year. That's the distance light travels in a year: about 9.4 trillion kilometres, or 63,000 AU. The disc of the galaxy is around 100,000 light years in diameter and 2,000 in thickness. The Solar System is located about 26,000 light years from the galactic centre, and takes something like 240 million years to complete an orbit around it.

So now we have a second diagram that's impossible to draw. Using a scale of one light year to the millimetre, the biggest thing from our previous diagram – the distance from the Sun to Proxima Centauri – is just 4.2 mm. In comparison, the diameter of the galaxy is a hundred metres – a football pitch – and our distance from its centre is 26 metres. Even the thickness of the disc, at two metres, is too big to fit on the page.

As it happens, it's that last measurement which may be the most relevant in the context of Oort Cloud perturbations. Within that two-metre disc thickness, the Solar System doesn't occupy a constant position. As it follows its huge orbit around the centre of the galaxy, it oscillates up and down on a somewhat faster timescale. Estimates vary, but it probably completes around four vertical oscillations per orbit. The amplitude of oscillation is between 300 and 600 light years – or between 0.3 and 0.6 metres on our imaginary diagram.

In the course of one of these oscillations, the gravity of the galactic disc – the cumulative effect of all the other stars in it – pulls first one way and then the other on the Oort Cloud. These alternating stresses might be enough to explain the influx of new comets into the inner Solar System, without the need to invoke close encounters with individual stars. This raises an intriguing possibility, too. If it is indeed the mechanism involved, the forcing effect would vary in a predictable, periodic way. Is the same true of cometary impacts?

Galactic nemesis

Various studies have hinted at a possible periodicity in both impact cratering and mass extinctions. The results, however, are far from conclusive, with a range of suggested periodicities between 25 and 35 million years. Some authors feel certain it's at one end of that scale, while others are equally certain it's at the other end. Whatever the exact number, there's no possible terrestrial – or even Solar System – mechanism that could produce a periodicity on such huge timescales. If it's real, its cause has to lie somewhere out there in the galaxy.

The existence of this sort of periodicity, and its possible link to galactic structure, was originally suggested in a paper by two British astrophysicists that appeared in the journal *Nature* in 1979. We've already met one of the authors – Bill Napier, who later went on to write the technothriller *Nemesis*. The other author was Victor Clube – a former Dean of Astrophysics at Oxford University, so we're not talking about a pair of cranks here (as some of their nastier critics have implied). In that *Nature* paper, Clube and Napier wrote that:

> The time sequence of terrestrial catastrophes and other Solar System phenomena is apparently stochastic but with an underlying galactic modulation.

'Stochastic' is a word you often see in professional scientific papers. It's almost exactly synonymous with the everyday word 'random', but scientists seem to prefer it for some reason. So all Clube and Napier are saying is that the distribution of impacts looks pretty random across time, but there's a subtle periodicity superimposed on it. Whether the second half of the statement is true is still open to debate, but there's no arguing with the first half. There will always be some random impacts, from asteroids and short-period comets, since the supposed periodicity is only going to affect long-period comets.

Although Clube and Napier championed the 'galactic oscillation' theory, that wasn't the only way to produce periodic impacts. An alternative – which is easier to understand, and hence more popular with the mass media – postulates the existence of a 'dark star' companion to the Sun, dubbed Nemesis.* The idea is that Nemesis remains undetected because it's very small and dim, but it's moving on an orbit which periodically takes it through the Oort Cloud. As a hypothesis, it's scientifically plausible – but not at all likely. Even its originators didn't take it terribly seriously. One of the first papers on the subject, by Marc Davis and collaborators in a 1984 issue of *Nature*, borders on the tongue-in-cheek:

> We propose that the periodic events are triggered by an unseen companion to the Sun, travelling in a moderately eccentric orbit, which at its closest approach (perihelion) passes through the Oort Cloud of comets which surrounds the Sun. During each passage this unseen solar companion perturbs the orbits of these comets, sending a large number of them (over one billion) into paths which reach the inner Solar System. Several of these hit the Earth, on average, in the following million years. At present the unseen companion should be approximately at its maximum

---

* There's no connection here to Bill Napier's novel of that title. The name 'Nemesis' is popular for looming threats because it originally belonged to the Greek goddess of retribution.

distance from the Sun, 2.4 light years, and it will present
no danger to the Earth until approximately AD 15,000,000.

The biggest problem, of course, is that a star that close to us
would probably be seen. If it was an ordinary star it would,
anyway. As it happens, however, there's more to the galaxy
than ordinary stars. The vertical oscillation we were talking
about earlier applies not just to the Sun, but to all visible
stars in the solar neighbourhood. It turns out that the total
kinetic energy of all that oscillatory motion is much too high,
if the only thing binding the stars to the galactic disc is their
own gravity. The stars would go up, and they'd never come
down again. Something else must be pulling them back.

The 'something else' is called dark matter, but we don't
know much about it. To be honest, we don't even know that
it exists – but if our current understanding of gravity is correct, then postulating its existence is really the only way to
explain the observed behaviour of visible matter. Beyond
that, dark matter could be almost anything – although there
are reasons for doubting that it's made up of objects as large
as Nemesis would need to be. On the other hand, there's
another way that dark matter could influence the Oort Cloud
– according to one theory, at least.

Because dark matter only interacts with ordinary matter
through gravity, most astronomers assume it only interacts
with itself in this way, too. That's different from ordinary
matter, which is subject to other, non-gravitational processes
such as the emission of energy in the form of heat and light.
These processes caused the normal matter to lose a lot of its
initial energy, resulting in the thin disc we see today. Unable

to follow suit, dark matter would have remained spread out in a much larger spherical halo around the galaxy. That's the standard theory, anyway.

A few years ago, a team led by Lisa Randall at Harvard University proposed an alternative picture. According to this theory, a fraction of the dark matter does indeed have its own private way of losing energy – and this allowed it to collapse down into a thin disc in the same plane as the visible disc. That, of course, would have the effect of boosting the gravitational perturbing effect on the Oort Cloud. According to Randall's team, only this model – rather than one involving ordinary matter or 'conventional' dark matter – is capable of providing a perturbation of the necessary magnitude.

As theories go, that's a pretty abstract one – certainly not the kind of thing you'd expect to generate screaming tabloid headlines. Yet that's exactly what did happen. Here's an example that appeared in the *Daily Mail* in March 2014: 'Were dinosaurs wiped out by DARK MATTER? Force sends comets hurtling towards Earth every 35 million years, claims theory.' Virtually identical stories appeared in many other media outlets around the same time.

It was great publicity for Randall's research (and for the book she wrote on the subject soon afterwards), but the dinosaur connection was tenuous at best. The Chicxulub impact that killed them off doesn't fit into her 35 million year cycle – and it may well have been an asteroid rather than a comet, anyway.

All this talk about periodicities and Nemesis and the galaxy is fascinating stuff – but it's a sideline to the main issue. The fact is that impacts occur, and even if there's a periodic element to them, remember what Clube and Napier said – the basic distribution is random (sorry, stochastic). So don't get a false sense of security because Nemesis is currently at aphelion, or because we're at the wrong point in the galactic oscillation cycle. A civilisation-destroying impact could happen at any time.

MAPPING THE THREAT   6

Arthur C. Clarke's 1973 novel *Rendezvous with Rama* is set in the 22nd century. It's not about cometary or asteroid impacts, it's about a huge alien spaceship – the eponymous Rama – that passes through the Solar System. For the story to play out the way he wanted it to, Rama had to be spotted as early as possible. The way astronomy worked in 1973, that probably wouldn't have happened. Big telescopes just weren't on the lookout for such things. Professional astronomers had a galactic fixation – not just with our own galaxy, but with the thousands of others beyond it as well. Solar System astronomy was passé, and searching for comets and asteroids was the unglamorous province of backyard amateurs.

So Clarke had to come up with a credible reason why the situation had changed by the time his novel was set, to ensure that Rama would be spotted as soon as it appeared. In a throwaway scene right at the start, a small asteroid scores a direct hit on the historic city of Venice, killing close to a million people. 'Project Spaceguard' is set up in the wake of this disaster, to keep constant watch on the skies for a similar threat in the future. As Clarke puts it: 'No meteorite large enough to cause catastrophe would ever again be allowed to breach the defences of Earth.'

In *Rendezvous with Rama*, this is merely the prelude to a story about an interstellar spaceship – one of the classic science fiction themes, of course. But it wasn't long before other sci-fi authors were employing impact scenarios in their own right. *Shiva Descending*, written in 1979 by astrophysicist Gregory Benford and William Rotsler, has already been mentioned. In this case, the threatening object is an Apollo asteroid called Shiva – a name which coincidentally (or otherwise) comes from the same Hindu pantheon as Rama. Clarke himself finally got in on the act in 1993 with *The Hammer of God* – this time featuring an Earth-threatening object named Kali, after yet another Hindu deity.*

By the time *The Hammer of God* appeared, the real world had caught up – as Clarke explains in an afterword to the novel. He proudly quotes the following passage from the International Near-Earth-Object Detection workshop that was hosted by NASA in 1991:

Concern over the cosmic impact hazard motivated the US Congress to request that NASA conduct a workshop to study ways to achieve a substantial acceleration in the discovery rate for near-Earth asteroids. This report outlines an international survey network of ground-based

---

* Pinning hazardous objects on one particular culture isn't a great idea, and hasn't been picked up in the real world.

telescopes that could increase the monthly discovery rate of such asteroids from a few to as many as a thousand. Such a programme would reduce the timescale required for a nearly complete census of large Earth-crossing asteroids from several centuries (at the current discovery rate) to about 25 years. We call this proposed survey programme the Spaceguard Survey (borrowing the name from the similar project suggested by science fiction author Arthur C. Clarke nearly 20 years ago in his novel *Rendezvous with Rama*).

Watching the skies

In the real world, Spaceguard isn't a single project but an umbrella term for various efforts around the globe to search for asteroids and comets that are likely to pass close to the Earth's orbit. Such objects are collectively known as near-Earth objects, or NEOs for short. It's an unimaginative name and a confusing acronym ('neo' usually refers to something new), but it's what they're called so we're stuck with it.

Although the hunt for NEOs is international, there's no getting away from the fact that its centre of gravity is firmly in the United States. NASA is by far the biggest source of funding, and all the largest and most successful surveys are US-based. Two of these are particularly noteworthy. First, there's the Catalina Sky Survey, which has been using two large telescopes in the clear air of the Catalina mountains near Tucson, Arizona since the end of the 20th century.

More recently, the Panoramic Survey Telescope and Rapid Response System – Pan-STARRS for short – became operational on the Hawaiian island of Maui in 2010. It was the latter team that found the interstellar intruder 'Oumuamua – which despite its distant origins still qualified as a NEO – in 2017.*

The basic techniques used to search for NEOs aren't too different from the ones developed by the celestial police at the beginning of the 19th century, when they were looking for a new planet and found the asteroid belt instead. The trick is to look at a small patch of sky, and then have a second look at the same patch a few hours later.

That's not quite as easy as it sounds, because the Earth is constantly rotating about its axis. If you simply left the telescope in the same position, before long it would be pointing in completely the wrong direction. Astronomers know that, though, so they keep the telescope moving in just the right way that it's always directed at the same point in space. If it's done correctly, the background stars won't move at all, because they're so far away. Anything that does move, on a timescale of a few hours, must be inside the Solar System.

In the days of the celestial police, when everything had to be done manually, this was a slow and tedious business. With modern computers and digital photography, however, it can all be automated, making the process thousands of

---

* After that discovery, and before the name 'Oumuamua was decided on, the *Economist* magazine made the following observation: 'Perhaps in expectation of a discovery like this, the International Astronomical Union, which approves such names officially, has not yet called an asteroid Rama. How about it, chaps?'

times faster – and much less soul-destroying for the humans involved. If it turns up a new object that's never been seen before, the next step is to share the details with the wider NEO community so that follow-up observations can be made.

Up to this point, the process will flag up anything that moves. It may be close to Earth, or it may be in the asteroid belt or the Kuiper belt. In the first case, it's obviously a NEO – but the converse isn't true. Even if it's a long way away when it's detected, the object may be travelling on a path that will bring it close to Earth at a later date. So there's a second stage to the process. Once a new object has been found, we need to work out where it's going.

If the object's motion were random, the problem would be next to impossible. All we see through a telescope is a projection of its three-dimensional motion onto the two-dimensional sky – and even that may have been observed for just a few hours or days. Fortunately, however, the motion isn't random at all. It obeys Kepler's laws, and that makes the problem tractable. Back in Chapter 2, we came across the idea of 'orbital elements' – five numerical parameters, such as perihelion distance and inclination to the ecliptic, which define an orbit completely. It turns out that just three observations, made at different times, are all you need for a preliminary orbit determination. This can then be refined as time goes on by adding further measurements.

Having said that, orbit determination isn't a simple matter – it involves some ferociously difficult mathematics. The problem first came to light in the wake of the discovery of Ceres in 1801. A few weeks after the dwarf planet was discovered there was a break in observations, and when they were resumed it couldn't be found again. Suddenly the astronomical world was desperate for a reliable method of orbit determination – and before long it had one, thanks to a young German mathematician named Carl Friedrich Gauss. As Carrie Nugent relates in her book *Asteroid Hunters*:

> Luckily for everyone, an honest-to-goodness mathematical genius was alive, and he thought the problem looked interesting. Carl Friedrich Gauss, then 24, turned his attention toward predicting where Ceres could be observed next. Accounting for the elliptical path and the movement of the Earth and Ceres relative to the Sun, Gauss formulated a complicated equation that he then solved using various approximation methods, some of which he invented on the spot.

The idea of solving an otherwise intractable equation using numerical approximations was a novel one in Gauss's day, and it wasn't until the advent of computers that it really came into its own. Some of the methods he invented for the Ceres problem are still used by software engineers today.

In the 20-plus years since astronomers started looking in earnest for NEOs, more and more of them have been discovered – some 18,000 by the end of 2017, as the following chart shows.

One of the most impressive successes came on 6 October 2008, when the Catalina team detected a NEO on a direct collision course for Earth, with impact predicted for the next day. As dramatic as this sounds, it was no cause for alarm

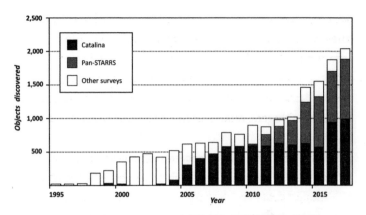

The number of near-Earth objects discovered each year from 1995 to 2017 (based on NASA data).

– the object in question, called 2008TC3, was only three metres across. Rather than posing a risk to the planet, it presented a huge opportunity for science.

The exact impact point was calculated, in the Sahara desert in north Africa. The locals were told to keep their eyes open, and the trail of the meteor as it entered the atmosphere was seen by observers in Egypt and Sudan. The object exploded in the upper atmosphere – but because it was so well tracked, people knew exactly where to search for rock fragments. Dozens were found over the coming months, totalling around 4 kg. It was the first time in history that a NEO had been seen and studied in three completely different ways – through telescopes as an object in space, in the sky as a meteor, and on the ground as meteorite fragments.

2008TC3 was one of the rare occasions (outside science fiction) that a NEO was discovered just hours before collision with Earth. Normally, even when there's a projected collision risk, it's years – or decades, or centuries – in the future. It's a matter of extrapolating the trajectories of both the offending object and the Earth, and seeing if they're ever going to be in the same place at the same time.

Surprisingly, it doesn't have to be *exactly* the same place. Imagine firing bullets at a circular target in a shooting gallery. If the bullet gets closer to the centre of the target than the radius of the circle, it's a hit – otherwise it's a miss. Now let's replace the target with the Earth, and the bullet with the NEO. Apart from the change of scale, is the situation exactly analogous? It turns out it isn't.

There's a good explanation of the problem in Bill Napier's novel *Nemesis* (which has been mentioned a couple of times already, and will feature again before the end of the book). Part of the novel's action centres around a group of NEO-hunters at a fictional observatory called Eagle Peak – located, like the Catalina sky survey, in the Arizona mountains. They're working on the assumption that an object called Nemesis is on a collision course with the Earth. Here's part of their discussion on the subject:

'Nemesis won't be going in a straight line. The Earth's gravity will curve it in,' objected Sacheverell.

Webb joined Schafer at the blackboard and they both started to scribble. Webb got there first. 'Hey, Herb finally got something right. Gravitational focusing will add to the Earth's target area. The gravitational target area exceeds the geometric one by V_E over V squared, where V_E is the escape velocity from Earth and V is the Nemesis approach speed.'

That sounds rather technical, but all it means is that a passing NEO is pulled in towards the Earth by the latter's gravity. In the shooting gallery analogy, it would be as if a bullet on a 'miss' trajectory suddenly curved in and hit the target. The algebra quoted by Napier's character means that the faster the NEO is going, the less it's pulled in. That's one advantage, from our point of view, of a high-speed encounter over a low-speed one. It all boils down to how that speed relates to the Earth's escape velocity.

As the name suggests, escape velocity – about 11 km/s – is the speed an object has to have to avoid being captured by the Earth's gravity. If a NEO comes in at this speed, then the effective target area is doubled – or increased even more if the speed is less than that. On the other hand, the gravitational focusing effect is reduced when the speed is increased. If it's doubled to 22 km/s, for example, the target area is only increased by a factor of one and a quarter – or just one and a ninth if it's trebled to 33 km/s.

If we knew the NEO's orbit precisely, the collision issue would be black and white. If the orbit passed through the target area it would be a hit, otherwise it would be a miss. In practice, however, it's never that simple. The calculated orbit is only an approximation, particularly in the period immediately after an object is discovered, when there are very few observations to go on. That means there's an uncertainty area surrounding the expected position of the object. For that reason, astronomers only ever talk about the 'probability' of collision, usually expressed as a percentage. It's essentially the size of the target area as a percentage of the surrounding uncertainty area.

There's an interesting but often misunderstood consequence to this. If the uncertainty area is large, but the Earth lies inside it, the probability of collision will be small but non-zero. As more observations are made, the uncertainty area starts to shrink. As long as the Earth remains inside it, this leads to an increase in collision probability – because you're dividing the same target area by a smaller uncertainty area.

With any luck, you'll get to a point where the uncertainty area has shrunk so much that the Earth is no longer inside it. When that happens, the probability suddenly drops all the way to zero. That's simple geometry – but it doesn't stop conspiracy theorists shouting about an establishment cover-up.

Something like that happened in a case that was mentioned a couple of chapters ago. When the asteroid Duende crossed the Earth's orbit in February 2013, astronomers knew it would come very close – but they also knew it couldn't possibly hit. Not everyone believed them – and some of the disbelievers felt vindicated when a huge meteor exploded in the skies above Chelyabinsk that day. As we've already seen, however, it was a different object altogether – one that astronomers didn't see coming.

The reason they didn't see it was that the Chelyabinsk asteroid came from the direction of the Sun. That meant it was only above the horizon in daylight, when telescopes couldn't see it (yes, it's night on the other side of the planet – but the telescopes there are looking in the opposite direction). This blind spot around the Sun is a constant irritation to NEO-hunters – they just have to hope anything hiding there comes out before it hits us. That didn't happen with Chelyabinsk.

In 2002, US Defence Secretary Donald Rumsfeld made a strange remark about 'known unknowns' and 'unknown unknowns', which many people wrote off as just another of those daft things politicians say. In fact, the distinction isn't as meaningless as it sounds. In the present context, an 'unknown unknown' would be something intrinsically undetectable, such as a black hole or a chunk of dark matter. Chelyabinsk, on the other hand, was a 'known unknown' – a perfectly normal asteroid hiding in a place we know that asteroids can hide.

There are other known unknowns too, dictated by the physical limitations of telescopes and shortcomings of the software used to sift through the search results. Nevertheless, the mere fact that such weaknesses are known gives us a certain power over them. We can, for example, count up how many 'known unknown' NEOs there are. Or to put it another way, we can work out what fraction of the total NEO population our current survey methods would miss. Based on this logic, astronomers calculated in 2012 that only 10 per cent of NEOs over 1 km in size remained undetected. That's a reassuringly low number – but they also worked out, more worryingly, that a massive 70 per cent of objects over 100 metres are still undetected.

Know your enemy

The most famous of all NEOs – qualifying as such because its orbit crosses our own – is Halley's comet. It was also the first NEO to be visited, by not one but five spacecraft, when it passed through the inner Solar System in 1986. The 'Halley Armada', as it was dubbed, consisted of two probes from Russia, two from Japan and one from the European Space Agency (ESA). The last, called Giotto, had by far the closest encounter – passing within 600 km of Halley's nucleus on 14 March. That wouldn't have been possible, however, without calibrating data provided by the other four probes, which flew past at greater distances a few days earlier.

As it was, Giotto was able to capture the first ever pictures of a cometary nucleus, unhidden by the surrounding dust and gas. It also saw the source of the latter, in the form of several distinct jets on the sunlit side of the comet. Chemical analysis of the ejected material showed that it was 80 per cent water, with the rest largely made up of carbon monoxide, methane and ammonia.

If this was a sci-fi movie, we could just say 'Giotto took off from Earth and flew to Halley's comet' and leave it at that. This being the real world, however, we need a bit more explanation than that. Let's start by looking back at the diagram of Halley's orbit on page 41. It crosses Earth's orbit in two places, where the Earth itself would be in October and May (in that order, because Halley goes round the Sun in the opposite direction to the Earth). During its 1986 passage, Halley reached perihelion in February, after which it headed back out towards its 'May' crossing of Earth's orbit. Earth was heading towards May too, of course, but not fast enough. Halley would get there in the middle of March.

It's clearer now what Giotto's task was. It had to find a short cut to the 'May' point of Earth's orbit, arriving there two months before Earth did – and at exactly the same time as Halley's comet.

For the nth time in this book, Kepler's laws come to our rescue once again. Space probes obey them just as fastidiously as everything else in the Solar System. The third law tells us that the time needed to complete an orbit is shorter for smaller orbits. This means Giotto could – quite literally – overtake Earth on the inside track. It was launched in July 1985 onto an orbit that was significantly more eccentric than Earth's. It had the same aphelion – 1 AU – but a perihelion much closer to the Sun, at 0.73 AU. By dropping down to that distance and then coming back out again, it managed to shave two months off its orbit – ending up in just the right place and right time for its close encounter with Halley.

It was a clever trick – but perhaps not quite as clever as it sounds. Giotto and the comet were still travelling very rapidly in different directions, so the encounter was a brief one. Giotto only had an hour or so before the diverging orbits took Halley out of range of its cameras. A cleverer trick – allowing a much longer, closer look – would involve the spacecraft matching orbits with the comet, and possibly even landing on it. Almost 30 years would pass before anyone pulled that trick off – and once again it was ESA, with its Rosetta-Philae mission.

As space spectaculars go, this was an impressive one – all the more so because it managed to capture public imagination in a way that no previous robotic space mission ever had. The day of the landing itself – 12 November 2014 – saw millions of tweets with the hashtag #CometLanding, making it one of the top trending topics worldwide. That was only partly down to the intrinsic science content of the mission – it was also helped by a particularly media-savvy public relations effort on the part of ESA, complete with cartoons featuring anthropomorphic versions of the Rosetta spacecraft and its Philae lander.

The target in this case was another short-period comet, called 67P/Churyumov–Gerasimenko. Its aphelion is at 5.7 AU, just beyond the orbit of Jupiter, and its perihelion at 1.2 AU, just outside the Earth's orbit. That makes it a NEO, though not a particularly hazardous one.

Rosetta was launched in March 2004, more than ten years before it finally reached comet 67P. That seems an inordinately long time; how can it possibly have taken so long to get there? Let's think it through. The furthest Rosetta could possibly need to travel – in fact the maximum distance from Earth it did reach – is the distance out to the orbit of Jupiter, or about 4 AU. But the Earth travels further than that in a single year.* So how did Rosetta manage to take so long?

There isn't a simple answer – it's a combination of several factors. First, there are Kepler's laws. Rosetta couldn't simply travel in straight line, it had to follow an elliptical path. The further it got from the Sun, the slower it moved round that ellipse. In the case of comet 67P, for example, although its perihelion is close to the Earth, it takes six times longer to complete an orbit because its aphelion is so much further out.

Rosetta was faced with another difficulty, too. It wasn't enough for it to get to the same place as 67P – it had to be

---

* Approximating the Earth's orbit as a circle of radius 1 AU, its circumference is 2π AU, or about 6.26 AU.

travelling on exactly the same orbit as the comet when it got there. That involved spiralling slowly outwards from the Earth's orbit towards that of 67P – hence the ridiculously long journey time.

When it finally got there, it had to reduce its relative speed to the point where it could enter orbit around the comet. It was at this point, however, that all those years of travelling started to pay off – as Rosetta sent back a series of spectacular images that were like nothing human eyes had ever seen.

Comet 67P/Churyumov–Gerasimenko, seen by the Rosetta spacecraft in September 2014.

ESA/Rosetta/NAVCAM, CC BY-SA IGO 3.0

Saying that Rosetta 'entered orbit around the comet' implies that the latter has its own gravitational field. That's true enough, as it is of any object possessing mass, but with a spatial extent of just 4 km it's not the strongest gravitational field in the Solar System. Even down on the surface of the comet, it's about a hundredth of one per cent of the Earth's gravity. That's so small you probably wouldn't even notice it.

This meant that getting the Philae lander down onto the comet – and making it stick there – wasn't an easy task. In fact, it didn't go quite according to plan, with the probe bouncing several times before finally ending up in deep shadow where its solar panels couldn't do their job. The lander had to fall back on a battery that could only supply 12 watts of power for three days. It had a lot to do – numerous experiments designed by different teams across Europe, all eagerly waiting for results – and there were other limitations to contend with too. On top of the ten-year trip out to the comet, the long lead time involved in space missions meant that Philae's electronics were 20 years behind the times. Amazing as it may seem, it only had a memory capacity of six megabytes.

As science writer Brian Clegg put it: 'Philae provided a resource-constrained, multi-project scheduling problem. Here the projects were the different experiments, while the resource constraints were the energy supply and the available memory.' Not surprisingly, there was frantic activity down on Earth to ensure that Philae could return the maximum amount of information in the limited time available. This involved a sophisticated mathematical technique called 'constraint programming', designed for just such a scheduling problem. To quote Emmanuel Hebrard, one of the people involved in the work:

> For each instrument there was a research team somewhere in Europe who designed it which was waiting on the results for about 20 years. Clearly, they all wanted enough resources for their experiment to run correctly, multiple times if possible. By carefully optimizing the activity plan, we wanted to help them do as much science as possible.

Fortunately, the scheme paid off, and Philae managed to carry out a chemical analysis of the comet's surface that would have been impossible any other way. It detected plenty of organic material (in the scientific sense described on page 82) – a discovery that was supported by further observations made by the main Rosetta spacecraft. This excited the interest of Chandra Wickramasinghe – who, following the death of his long-time collaborator Fred Hoyle, was now the world's leading proponent of the 'extraterrestrial microorganisms' theory. As Wickramasinghe put it:

> If the Rosetta orbiter has found evidence of life on the comet, it would be a fitting tribute to mark the centenary of the birth of Sir Fred Hoyle, one of the undisputable pioneers of astrobiology.

Most scientists were much more cautious – including ESA's own mission team. When asked about 'evidence of life on the comet', they brushed the suggestion off as 'pure speculation' and 'highly unlikely'.

As high profile as the Rosetta mission was, it wasn't completely without precedent. Even before it was launched, NASA's Near Earth Asteroid Rendezvous (NEAR) spacecraft had performed a similar feat with Eros – an Amor-type asteroid with perihelion just outside Earth orbit, at 1.1 AU. At more than 30 km across, it's one of the largest of the known NEOs. Yet there was no great media fanfare when NEAR went into orbit around Eros in February 2000, and proceeded to land on it a year later.

Perhaps asteroids don't hold the same fascination for the public as comets do – or maybe it's just that NASA's PR skills were no match for ESA's. It didn't help the spacecraft's image to burden it with a clunky name like NEAR (or NEAR-Shoemaker, as it became after the death of Eugene Shoemaker, one of the co-discoverers of comet Shoemaker-Levy 9). In terms of achievement, however, the mission was every bit as impressive as Rosetta. Unlike the glitch with Philae, the landing went flawlessly, and the spacecraft continued to return useful data for more than two weeks afterwards.

Unfortunately, that data merely confirmed what everyone already suspected: Eros is just a big lump of rock. There are no volatiles to make a dramatic tail, and no amino acids to raise speculations about extraterrestrial life. As NASA's own website unglamorously puts it, Eros is 'a solid, undifferentiated, primitive relic from the Solar System's formation'.

That's not to say that all asteroids are boring. Some of

them are actually quite interesting – the kind that carbonaceous meteorites come from, for example. There's one in that category called Bennu, which happens to be an Apollo asteroid on a potentially hazardous Earth-crossing orbit. It's the target for NASA's next asteroid mission – which they've given the more dramatic and imaginative name of Osiris-Rex. That's an acronym for something or other, but Osiris is also the name of the Egyptian god of the dead. It's a suitably ominous choice in the present context, because Bennu has a small but measurable chance of hitting the Earth at some point in the future. With a diameter around 500 metres, that would mean a 15,000 MT explosion.

The goal of the Osiris-Rex mission is to pick up a rock sample from Bennu and bring it back to Earth. That won't be a first in itself; as long ago as 2010 the Japanese spacecraft Hayabusa came back with a sample from the Apollo asteroid Itokawa; a follow-up, Hayabusa 2, is aiming to duplicate the feat with another near-Earth asteroid, Ryugu.* The difference with Osiris-Rex is that the hoped-for sample will be much larger – potentially a kilogram or more, compared with less than a gram. Having such a sizeable sample of an asteroid will allow a much more thorough analysis of its composition – and, after all, it's always best to know your enemy.

---

* At the time of writing (late 2018), Hayabusa 2 is in orbit around Ryugu.

Asteroid Mining

It's already been mentioned, in the context of the Chicxulub impact (see page 63), that asteroids contain the heavy element iridium which is extremely rare in the Earth's crust. That's not just of academic interest; iridium is an industrially valuable material used in the manufacture of electronic components. The same is true of several other elements that are easier to find in asteroids than near the surface of the Earth, such as osmium, palladium and platinum. The latter is often lumped with gold as a substance whose monetary value lies solely in its scarcity, but that's not true. Platinum is a key ingredient in several modern technologies – and Earth's supply of it could be depleted in a matter of decades.

If NASA's Osiris-Rex mission is successful, it will demonstrate the feasibility of bringing asteroid material back to Earth. A number of private companies – with exciting names like Deep Space Industries and Planetary Resources – have plans to develop asteroid mining in the not-too-distant future. Such activities would have a secondary use too. As well as the rare elements already mentioned, asteroids contain materials that are common on Earth but costly to get from Earth into space – such as iron, aluminium and even water – which could be used as raw materials for space construction projects.

So asteroids aren't just a space hazard – they're an opportunity too. In a later chapter we'll look at some of the options for deflecting asteroids away from a collision course with Earth. Hopefully, the requisite hardware wouldn't have to be used very often – so would it even get built? It's more likely to see the light of day if it had a dual role in asteroid mining.

Assessing the risk

In Britain, one of the first high-profile figures to talk about the impact threat was the politician Lembit Opik. In 1999, the BBC quoted him as saying:

> I'm calling on the government to take seriously the pros-
> pect of asteroid or cometary impact with the Earth. Now
> that's got a pretty high giggle factor, it makes me sound
> like one of those millennium soothsayers, a Nostradamus
> of parliament, but actually it's a very serious threat.

As regards convincing the government, or the public, of the seriousness of the threat, Opik may not have been the best person for the job. With a reputation as one of the country's more eccentric politicians, he had something of a 'giggle factor' of his own (after losing his seat in parliament, he even dabbled in stand-up comedy for a time). Nevertheless, Opik was qualified in a way. His grandfather, Ernst Opik – a native of Estonia – had been a professional astrophysicist. In 1932, anticipating Oort by almost 20 years, he suggested that long-period comets might originate in a spherical cloud on the outer edge of the Solar System. He was also one of the first astronomers to take the threat of asteroid collision seriously.

In 1958, a paper by the elder Opik called 'On the cata-strophic effects of collisions with celestial bodies' appeared in the *Irish Astronomical Journal* (he was working at Armagh Observatory in Northern Ireland at the time). Based on a sta-tistical study of asteroid sizes, he estimated that on average a one-kilometre object would hit Earth every three million years. This figure has nothing to do with the periodicity we were talking about in the previous chapter – it's just a consequence of random chance.

Opik also noted that, because smaller objects are more numerous than large ones, they're also likely to hit us more often. He estimated that the time interval drops to 600,000 years for a half-kilometre object, 100,000 years for 250 metres, and 10,000 years for 100 metres.

This was an important insight, which was decades ahead of its time. The world had to wait more than 50 years for an official view on the subject. It eventually came in 2010, in a report by the US National Academy of Sciences dramatically titled *Defending Planet Earth* (its subtitle was the more prosaic 'Near-Earth object surveys and hazard mitigation strategies'). It includes a similar analysis to Opik's – but in more depth, and using up-to-date statistics on the numbers of NEOs. The results are presented in graphical form in the report, but in round numbers they can be summarised as follows:

Estimate of impact frequencies for objects of different size

Object size (in metres)	Estimated number in near-Earth orbits	Years between impacts, on average	Energy in megatons
20 (Chelyabinsk-sized)	10,000,000	80	1
60 (Tunguska-sized)	1,000,000	400	10
100	100,000	5,000	100
1,000	1,000	500,000	100,000
10,000 (Chicxulub-sized)	10	50,000,000	100,000,000

It's noticeable that the new study gives systematically higher impact frequencies than Opik – a consequence of the fact that far more NEOs are known today. One thing hasn't changed, though – these are still meant to be interpreted as statistical averages, not a periodic pattern. Think of the analogy of weather statistics. A London guidebook might tell you that it rains there on one day in three on average. No one's going to interpret that to mean that if it rains one day, it's guaranteed to be dry for the next two days. It's the same with impacts. Just because a 60-metre asteroid hit Tunguska a hundred years ago, and the table says there's an average gap of 400 years between such impacts, that doesn't mean we're safe for the next 300 years.

Looking back at the table, two obvious facts jump out: firstly, that small impacts are much more likely than large ones; secondly, that large impacts are much more destructive than small ones. So which should we worry about more – an unlikely event that would cause global devastation, or a more probable one that might only wipe out a single city? The truth is, we can't ignore either possibility.

An effort to put some order into the situation was made in 1999, at a conference in the Italian city of Turin. That's what the English-speaking world calls it, anyway – but its Italian name is Torino, so the thing that came out of it is called the Torino scale. It's an attempt to quantify, on a scale of 1 to 10, the hazard associated with any impact that's projected to occur less than a hundred years in the future. The scale isn't as simple as it might be, because there are two variables to consider – the size of the impactor and the probability of it hitting the Earth.

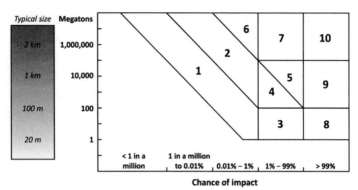

The hazard level on the Torino scale depends on the size and probability of impact (note that both scales are nonlinear).

The easiest way to see how the Torino scale works is to put a specific example into it. For a 60-metre, Tunguska-sized object the hazard level would be 1 if an impact was possible but unlikely, 3 if there was a reasonable chance of an impact, and 8 if it was virtually certain.

Now for some good news – there's currently (at the time of writing) no known object with a Torino rating even as high as 1. In other words, everything we know about is either too small to cause any damage on the ground, or there's zero chance that it will collide with Earth in the next hundred years.

That hasn't always been the case. Over the years since its inception, dozens of new NEOs have been flagged at level 1 on the Torino scale when they were first discovered, only to be downgraded after their orbits had been computed with greater accuracy. Only one object has ever scored higher than level 1, and that was a rather worrying 4. It was an Aten-type asteroid called Apophis – at 350 metres, much larger than Tunguska and well into the thousand-megaton class.

You may remember that Aten-type asteroids, with an aphelion just outside the Earth's orbit and a perihelion well inside it, manage to complete a circuit of the Sun in less than a year. In the case of Apophis, the orbital period is just 10 months, giving it plenty of opportunities to come close to the Earth. When it was first discovered in 2004, it looked like there might be an impact in 2029, with a probability of just over 1 per cent. Combined with its 350 metre size, that's how it acquired its Torino level of 4.

Fortunately, subsequent observations refined the orbit and ruled out a collision – so Apophis dropped right off the Torino scale. It will still make a very close pass in 2029 – much closer than the Moon, and possibly even visible as a faint star-like object to the naked eye – but there's no risk it will hit Earth.

That gives us time to think what we could have done about it if it really had been on a collision course.

PLANETARY DEFENCE 7

The idea of launching a space mission to avert a cosmic collision entered public consciousness in 1998, with not one but two Hollywood films on the topic. First, in May that year, came the atrocious *Deep Impact* – full of scientific implausibilities and with an unbearably slow-moving plot. If nothing else, it's proof that there really is such a thing as a boring disaster movie. That couldn't be said of *Armageddon*, released two months later, which is packed with wall-to-wall, edge-of-the-seat excitement. On the science side, however, it makes *Deep Impact* look like a PhD thesis. According to *Armageddon*'s trivia page on the Internet Movie Database:

> NASA shows this film during their management training programme. New managers are given the task of trying to spot as many errors as possible. At least 168 have been found.

That's probably apocryphal – but it might just as well be true. They're not small errors, either – they're gigantic, ludicrous ones. In one scene, the incoming asteroid is described as 'the size of Texas'. In other words, it's about 1,200 km across – a hundred times the size of the dinosaur-killing Chicxulub impactor. Well, sorry – but there just isn't an asteroid that big. The largest object in the asteroid belt – the dwarf planet Ceres – falls quite a way short at just under 1,000 km.

And it gets worse. We're told, with just 15 days to go before impact, that 'there are only nine telescopes in the world that can spot the asteroid'. That just doesn't stand up. You only need to do a few simple calculations to see that, if the asteroid really was as big as they say, it would be easily visible to the naked eye by that time.

Those are just some of the scientific errors in *Armageddon* and there are many more. But is that really such a terrible thing? *Deep Impact* and *Armageddon* may be bad films, but they did a lot to bring the collision threat to public attention. All in all, the positives probably outweigh the negatives.

Of course, there's another thing about those two movies. They conveyed the impression that we could destroy an incoming asteroid or comet, using present-day nuclear weapons. Is that true – or just another of the facts they got wrong?

How to blow up an asteroid

Forget an asteroid the size of Texas – or even the more sensible, but still highly unlikely, 11-km comet featured in *Deep Impact*. The latter film is allegedly based on Arthur C. Clarke's novel *The Hammer of God*, although any similarities that might exist between the two are difficult to spot. *The Hammer of God* isn't Clarke at his best, but at least it makes a decent effort at scientific credibility. In the novel, as mentioned previously, the threat comes from an object named Kali, which could qualify as either an Apollo asteroid or an aging comet. Its size ('1,295 metres maximum length, 656 metres minimum width; Kali would fit easily into many city parks') is much smaller than its Hollywood cousins, but still large enough to cause enormous destruction if it hit Earth. So let's focus on that for a moment. Could we blow up an object the size of the fictional Kali with present-day nuclear weapons? Let's think it through.

The most powerful nuclear weapons are the hydrogen bombs that proliferated during the Cold War, and still exist in large numbers today. The first American test of such a weapon, codenamed Ivy Mike, took place on 1 November 1952 on the island of Elugelab in the Pacific Ocean. Now, if you look up Elugelab on Wikipedia, you'll see that it speaks of the island in the past tense. That's because it ceased to exist on 1 November 1952. The Ivy Mike explosion produced a crater almost 2 km in diameter – which just happened to be bigger than the island itself.

In the same way, a 2-km crater on an asteroid 1.3 km across would be pretty terminal for the asteroid. Ivy Mike was about ten megatons – typical of a Cold War nuclear weapon – and there's nothing to stop you exploding several warheads of that size if you wanted to make absolutely sure. So there are no two ways about it: you really can blow up an asteroid with a nuclear explosion.

That's not to say it's a good idea, though. The fragments – some large enough to do a considerable amount of damage

– would continue to follow the same course towards Earth. That's a point that was made in a short story called 'How to Blow up an Asteroid', by the amateur astronomer and science populariser Duncan Lunan. Dating from 1973 – long before the subject became fashionable, and hence almost completely ignored at the time – the story describes the use of nuclear missiles against a 1.5-km asteroid on a collision course with Earth. Although it's blown to pieces, the chunks still continue on their way and hit Earth. The result is a lot of small impacts, instead of a single big one.

For this reason, the 'blow it up' option – or the disruption option, as scientists call it – is a long way down the list of solutions to the cosmic impact problem. That's not because it's impossible, or even particularly difficult – it's just not the best way to go about things. The smart money is on 'deflection' rather than 'disruption'. To see why that is, we need to talk about orbits again.

The problem with blowing something up is that the resulting fragments are still going to follow pretty much the same orbit. The fact that they were blown apart with different speeds means the orbits aren't precisely identical, but it will be a long time before they diverge enough to miss the Earth. Instead of blowing the object up, what we really need to do is move it out of harm's way.

On its own, the fact that an orbit intersects the Earth's isn't a big problem. There are thousands of NEOs on such orbits, any one of which would cause massive devastation in the event of a collision. But that requires something else besides an orbit crossing. Both the Earth and the NEO have to be in the same place at the same time. From the NEO's point of view, it's a question of hitting a moving target – and a really fast moving one, at that. There are far more ways that a NEO can miss Earth than hit it.

For the same reason, if a collision is about to happen, there are many ways the orbit could be changed to avert it. The obvious one – pushing the object sideways so it misses – is only one of the options. A less obvious one is to slow it down or speed it up a little. With the Earth whizzing round the Sun at 30 km/s, it only takes the planet a little over seven minutes to move a distance equivalent to its own diameter. If the NEO gets to the intersection point slightly too early or slightly too late, we're going to be safe.

Things are starting to look easier now. All we need to do is give the NEO a push – it really doesn't matter which way. If we push it sideways, it will miss. If we push it backwards, it will slow down – and miss. If we push it forwards, it will speed up and miss.

Of course, you know there's going to be a catch – and here it is. That push has got to be a very big one indeed. A NEO large enough for us to worry about will be much more massive than anything we're used to pushing around – like an oil tanker or an aircraft carrier. A one-kilometre rock amounts to more than a billion tons – or something like 10,000 aircraft carriers. How are we going to push that into a new orbit, even if it's only by a small amount?

Actually, a nuclear explosion is still a pretty good option here. One of the first proponents of this idea was the physicist Edward Teller. Like Harold Urey and Luis Alvarez, who cropped up in an earlier chapter, Teller was one of the scientists who worked on the Manhattan Project to design the first atomic bomb. After the war, he played a major role in creating an even more destructive weapon – the multi-megaton hydrogen bomb. That was far from a single-handed achievement, but it gained Teller a reputation as 'the father of the H-bomb'.

That's an accolade most people would be ashamed of – but Teller was proud of it. For the rest of his life, he advocated the use of H-bombs in any context he could think of – and not just military ones. In the 1960s, based on its tried-and-tested ability to create a big crater in the ground, Teller suggested that an H-bomb would be the perfect tool to dig out an artificial harbour. Then in 1995, at the age of 87, he turned up at a Planetary Defence Workshop with a new take on the 'blowing up an asteroid' theme.

Teller proposed that a small section of the asteroid's surface should be chopped up into rubble by some mechanical means, after which – in his own words:

> We can then put a nuclear explosive close to the surface, which will irradiate the rubble that we have already created. This tends to homogenise this rubble and push it one way, while, by reaction, the remaining 90 per cent of the material is pushed the other way. The reaction on the main body will be very powerful, and there can be no doubt that appropriate deflections can be arranged.

It's not a stupid idea – but it's a complicated one. Teller's method involves spending a considerable length of time working on the asteroid, which would require a spacecraft to match speeds with the asteroid and land on it. As we saw in the last chapter, that's a process that can take years. It would be much quicker and simpler if the whole thing could be done in a single, quick pass (compare Giotto's eight-month mission to fly past Halley's comet with the ten years it took Rosetta to match orbits with comet 67P).

All we really need to do is get close enough, for as brief a time as it takes to detonate the bomb, to produce something called a standoff explosion. After that, the laws of physics do the rest – as Gerrit Verschuur explains:

> The explosion creates so much energy, in particular in the form of fast-moving neutrons, that the surface of the aster-oid on the side of the explosion is heated slightly and gas and dust begin to stream away from the surface to act like a jet that pushes the asteroid into a different orbit.

Exactly this approach is employed in Gregory Benford and William Rotsler's novel *Shiva Descending*, referred to earlier. It's arguably the best science-fictional treatment of the asteroid deflection problem, certainly in terms of getting the physics right (which isn't that surprising since, as noted elsewhere, Benford was a professor of astrophysics). One of the novel's characters makes the interesting point that, no matter how powerful the bomb is, only a tiny fraction of its energy actually goes into changing the course of the asteroid (called Shiva in the story):

> Compared with simply converting the burst energy into a change in Shiva's kinetic energy, the bomb is about 3 per cent efficient.

Of course, that's only one of the arguments against this particular approach – there's also the inevitable controversy that surrounds all things nuclear. More than a hundred countries have ratified the Outer Space Treaty of 1967, which includes an undertaking 'not to place in orbit around the Earth any objects carrying nuclear weapons'. That clause could be disputed in the context of an asteroid deflection mission – when the bomb wouldn't technically be a 'weapon', and it would be in orbit around the Sun rather than the Earth – but that's really only playing with words. In any case, the Comprehensive Nuclear Test Ban Treaty of 1996 prohibits all nuclear explosions in space, whether for military or civilian purposes.

So let's forget the nuclear option for a moment, and see if we can find a more politically acceptable – not to mention energy-efficient – way to divert a NEO off a collision course.

Hitting a bullet with a bullet

The most efficient way to transfer energy to a NEO is to crash into it at high speed. That doesn't require any kind of warhead – just precision aiming at an eye-wateringly high closing speed. It's a bit like hitting a bullet with a bullet – an image that was often used during the Cold War to describe the anti-ballistic missile (ABM) problem. The first idea was simply to destroy an incoming nuclear missile with another nuclear missile – which didn't have to get particularly close to its target, because the radius of destruction was so high. Things became a lot trickier after nuclear-tipped ABMs were banned by international treaty in 1972. Now it was a question of choreographing a head-on collision between an unarmed ABM and the incoming missile, and hoping that kinetic energy alone would be enough to destroy it. That called for pinpoint accuracy in both space and time.

Intercepting a NEO presents all the same challenges as the ABM problem – plus another one. In the missile-on-missile scenario, the two objects are going to be fairly evenly matched in terms of their mass and speed – and hence their kinetic energy. There's no such balance in the NEO case, however, where the target may have a billion times more kinetic energy than the interceptor.

As difficult as the problem is, it's one that's taken seriously by the science community. NASA calls it a kinetic impactor – and it even tried something like it once, on a very small scale. In 2005, comet Tempel 1 had a visitor in the form of a spacecraft called Deep Impact. The name was a nod to the dire 1998 movie of that name – but not because Tempel 1 posed any impact threat to the Earth (it's not even in a near-Earth orbit). The recipient of the impact on this occasion was the comet itself – as NEO-hunter Carrie Nugent explains:

> The goal of that mission was to study the surface by making a crater and stirring up the surface material so it could be studied. To do that, the spacecraft hit the comet with something heavy: a 372 kg probe travelling at 10 km/s. That's faster than a bullet. The mission's goal wasn't to change the comet's orbit, but as you might expect, the orbit

did change, very slightly. The impact altered the comet's velocity by about 0.00005 mm/s [millimetres per second].

Comet Tempel 1 was already covered in impact craters – and now it has one more, thanks to NASA's Deep Impact probe.

Up to a point, it all looks very promising. Deep Impact took less than six months from launch to arrival at the comet. That's a lot faster than Rosetta's encounter with comet 67P, because the aims of the two missions were completely different. Rosetta wanted to match the comet's speed as closely as possible, while Deep Impact wanted to do the exact opposite. The faster it was moving relative to the comet, the higher the kinetic energy of its impactor probe would be.

Even though Deep Impact wasn't trying to change the comet's orbit, its probe hit with the right sort of speed to do so – 10 km/s. The reason it produced so little effect was its tiny mass – the other factor in kinetic energy besides speed. To a first approximation, the impactor simply transferred its momentum – mass times velocity – to the comet. On that basis, if the comet had a trillion times the mass of the

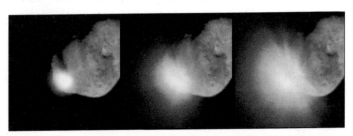

**Three video frames showing the Deep Impact
probe hitting comet Tempel 1.**

NASA image

impactor, the change in its speed would be a trillionth of the impactor's speed.

It's actually better than that, by a factor that scientists call beta. This comes into play because the impact blasts a large amount of the comet's own material into space – as you can see in the NASA images – and this carries off additional momentum. The value of beta varies between one* and ten, depending on the composition of the object. If it's dense and hard, beta is high; if it's spongy and porous, beta is low.

Of course, if NASA's Deep Impact team had really been trying to alter the course of Tempel 1, they would have used a much more massive impactor – or a whole fleet of them – and probably a higher closing speed too. What's more, they wouldn't just try something and hope it worked – they'd do calculations in advance to make sure that both the mass and speed were high enough to have the desired effect.

This is the sort of calculation scientists love doing. Give them an Earth-threatening object, of a particular mass and on a particular orbit, and they'll work out exactly how hard you have to hit it to turn a collision into a comfortable miss. A theoretical study of exactly this type appeared in the journal *Acta Astronautica* in February 2018, under the title 'Options and uncertainties in planetary defence: Mission planning and vehicle design for flexible response'. It had no fewer than 15 authors from various US government agencies – including six from NASA.

That last fact, coupled with the theoretical concept's

---

* In theory beta could be less than one – or even negative – if you were unlucky enough to trigger a massive eruption of material on the opposite side from the impact.

eyecatching name of HAMMER (for Hypervelocity Asteroid Mitigation Mission for Emergency Response), led the tabloid press to get the wrong end of the stick altogether. 'NASA building HAMMER spacecraft to save Earth from cataclysmic asteroid impact,' screamed a headline in the *Daily Mirror*.

In reality, no one is building anything – that would require a team of engineers and a multi-billion dollar budget. All that we have here is a group of theoretical scientists doing some mathematical calculations. That costs very little money, it's what scientists do all the time, and it really isn't the stuff of shouty tabloid headlines. If you want to know the real facts, the *Acta Astronautica* article is freely available online – and it makes interesting reading.

As a case study, the scientists picked the asteroid Bennu – the destination for NASA's Osiris-Rex mission. On the basis of current orbit calculations, Bennu will make several close approaches to the Earth in future, but nothing close enough to constitute a collision risk. That, of course, was no use at all from the study's point of view – so the authors pre-tended that Bennu's orbit was slightly different, and it was going to hit Earth in 2135. Their problem then was to work out what sort of kick it would take to prevent that impact.

The first question is how much Bennu's speed needs to be changed in order for it to miss Earth. As a general rule, shifting its orbit by one Earth radius ought to do the trick (that's enough to change a dead-centre hit to a grazing miss). The effort needed to achieve this depends on when that speed change is applied. It turns out that, in very rough terms, it's 200 mm/s divided by the number of years to go to 2135. So if you give it a kick in 2125, you have to change its speed by 20 mm/s, but if you manage to get in there 25 years ahead of time, in 2110, you only need 8 mm/s. The latter figure, according to the study's authors, lies within the realm of possibility. Assuming a value of 2.5 for beta, they calculate that the desired effect could be produced by a salvo of three HAMMER interceptors launched by NASA's next-generation Space Launch System.

For a well-studied NEO like Bennu, a 25-year lead time shouldn't be a problem. In such cases, in fact, we'd probably have a century or more's notice of an upcoming collision. On the other hand, for an Oort-cloud comet that wasn't detected until it reached the orbit of Saturn or Jupiter, there'd be a lot less thinking time – possibly only a year or two. The HAMMER team has an answer to that one, too – and it's an answer the sci-fi fans are going to like: 'If the time to impact is too short for a kinetic impactor mission to deflect the NEO … then robust disruption via nuclear device becomes the last alternative.'

Fortunately, with the current focus on finding and tracking as many NEOs as possible, short warning times are going to get rarer in future – and long ones are going to get even longer. With warning times potentially measured in centuries, that opens up a whole new approach to NEO deflection.

The gentle touch

As mentioned in the previous chapter, when the National Academy of Sciences produced a report on *Defending Planet Earth* in 2010, they gave it the plodding subtitle 'Near-Earth object surveys and hazard mitigation strategies'. Under the heading of 'mitigation', the report talks about nuclear explosions and kinetic impactors – but it also describes a much gentler approach that it calls 'slow-push-pull'. It defines this as follows:

> 'Slow-push-pull' means the continuous application of a small but steady force to the NEO, thereby causing a small acceleration of the body relative to its nominal orbit. The effect of such small accelerations is most productive if applied along or against the NEO's direction of motion, as this causes a net shift of the NEO along its orbit. This shift can avert an impact by causing the NEO to show up at Earth's orbit earlier or later than Earth does.

From the point of view of a space mission, slow-push-pull is trickier than the nuclear or kinetic options – which is probably why it wasn't considered in the HAMMER study – because the spacecraft has to match speeds exactly with the NEO. As we saw in the case of Rosetta, it can take ten years to do that. But this was never meant to be a quick solution – that's why it's got 'slow' in its name.

The first option the report considers is a 'tugboat spacecraft to physically push on the NEO, similar to a tugboat moving a much larger ship by applying a small but consistent force'. That sounds fair enough – but there's a catch. Contrary to the impression given by Hollywood movies, spacecraft aren't under power most of the time – they just coast along at a constant speed. In fact, because of their limited supply of fuel, conventional rocket motors can only be used in short bursts lasting a few minutes at most. That's no use to us here at all – what we need for slow-push-pull is something that can apply a continuous thrust over a period of months or years.

A potential solution, capable of providing both the necessary endurance and a decent amount of thrust, would be a rocket powered by nuclear fusion – the same type of nuclear reaction that takes place inside the Sun. At the moment there's no such thing as a fusion rocket, but it doesn't break any laws of physics so it's something that might exist in the future. That's the assumption that Arthur C. Clarke makes in *The Hammer of God* – where Plan A involves attaching fusion rockets to the offending space rock, Kali, and altering its orbit that way (Plan B is that old sci-fi favourite – a salvo of nuclear missiles).

Clarke calls his fusion rocket Atlas – not a particularly original name for a rocket (there's already a whole family of them), but an etymologically sound choice in this case:

> The task of the mythological Atlas was to stop the heavens crashing down upon Earth. That of the Atlas propulsion module … was much simpler. It had merely to hold back a very small portion of the sky. … It could run continuously not for mere minutes but for weeks. Even so, its effect on a body the size of Kali would be trivial – a velocity change of a few centimetres per second. But that should be sufficient, if all went well.

Even a nuclear rocket has to obey Newton's law about action and reaction being equal and opposite. It can't provide thrust unless it has a constant supply of reaction mass. The obvious source of this, although Clarke doesn't use it in his novel,* would be to scrape off some of the NEO's own material. That shouldn't be too difficult, because both asteroids and comets have surfaces that are quite crumbly and soil-like.

That crumbly nature, however, also causes problems for this particular approach to slow-push-pull. It makes it difficult to get a really solid grip on the thing you're pushing and/ or pulling (or, in the case of an asteroid of the 'rubble pile' type, downright impossible). There's another complication too. Most NEOs are rotating – and you'd have to find some way of halting that rotation before the method would work.

There's a much simpler alternative, which doesn't require any physical contact with the NEO at all. It goes by the name of 'gravity tractor' – and it's much closer in concept to a *Star Trek* tractor beam than a farm tractor, since the latter needs to be physically attached to the object it's towing. A gravity tractor, on the other hand, uses an invisible beam of energy like its sci-fi counterpart. Well, okay – it uses the force of gravity, but that's pretty much the same thing.

It exploits the fact that there's a gravitational force of attraction between any two bodies – and that includes a NEO and a spacecraft flying alongside it. The force has a more noticeable effect on the spacecraft, because it's the smaller of the two objects – but it works the other way, too. Over the

---

* Eschewing this obvious approach, Clarke has his Atlas rocket take its own self-contained supply of reaction mass along with it in the form of liquid hydrogen. Why do it that way? A cynic would say it's because he wanted to include a dramatic scene in which someone sabotages the hydrogen tanks.

course of years, the spacecraft will slowly but surely pull the NEO out of harm's way. The spacecraft would need to apply a constant thrust to prevent it from crashing into the NEO, but this would be tiny compared to the thrust needed in the 'tugboat' concept, because the force on the NEO itself now comes from the spacecraft's gravity.

However, for the types of situation considered up to this point, talking about achieving the necessary effect 'over the course of years' is hopelessly optimistic. If the aim is to shift the NEO's orbit by an Earth radius or more, the tiny force produced by the gravity tractor would take centuries. But there's another situation where only a small deflection is needed to save the day – and that's where the gravity tractor really comes into its own. Let's run through a specific example.

The case of Apophis was mentioned in the previous chapter. That's the Aten-type asteroid that briefly peaked at level 4 on the Torino scale, with a potential collision risk in 2029. When that original orbit calculation was done, another possibility came up – that Apophis would miss in 2029, but come close enough to Earth on that occasion that the planet's gravity would deflect the asteroid onto a new orbit. This altered orbit would then produce a collision seven years later, in 2036. The whole thing was extremely finely balanced. In order for it to happen, Apophis had to pass through a small 'keyhole' in space that was less than a kilometre across.

Now that Apophis has been tracked for a longer time, we know it's not going to come close to the keyhole in 2029. But that was just one example – Apophis might still hit another keyhole at a later date, or a similar situation might arise with another asteroid. If it did happen, then a gravity tractor

– pulling steadily on the asteroid over a period of several years – might be all that was needed to shift it away from the keyhole.

There's another slow and steady way to alter an asteroid's orbit – and it's one that nature invented. It's called the Yarkovsky effect, after the Russian engineer Ivan Yarkovsky who predicted it way back in 1902. It's all to do with the curious behaviour of an asteroid in the night-time.

It's already been mentioned that most asteroids rotate, and this gives them a cycle of day and night just like the Earth. The daylight side of the asteroid absorbs sunlight, while the night side radiates it back out into space. In physics terms, the asteroid's night side is constantly emitting a stream of photons, which carry off a small amount of momentum just like a tiny rocket thruster. Given enough time, that's enough to change the asteroid's orbit. For Yarkovsky, this was just a hypothesis – but we now know it's a perfectly real effect. It's been measured in the case of Osiris-Rex's asteroid, Bennu.

The size of the Yarkovsky effect depends on the physical properties of the asteroid, and on its surface brightness in particular. That gives us a very simple way to control it artificially – just paint the rock a darker or lighter colour, as needed. Given enough time, that might even allow us to shift an asteroid all the way from an Earth-crossing orbit to a non-Earth-crossing one.

Finally, there is one more potential course of action which should not be overlooked. With a Spaceguard-type system in place to give us advance warning of incoming threats, it may not always require a space mission to deal with them effectively. For objects below a certain size, when the effects are likely to be local rather than global, it would be simpler to identify the impact point as accurately as possible and organise an evacuation of the area. That would be the best course of action, for example, in the case of a Tunguska-sized object, or the one that created the Barringer crater in Arizona.

So we've ended up with a range of options to choose from. Which of them is best will depend on the size of the object and the amount of warning time before a collision. The following diagram summarises the possibilities (without making any attempt at numerical quantification, because there are so many variables and uncertainties):

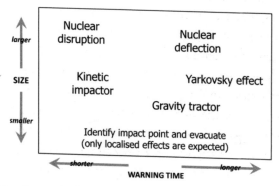

The best options for dealing with a NEO threat will depend on its size and the amount of warning time.

It's clear that anything but the smallest of threats is going to call for a space mission. Whether that could be mobilised on the necessary timescale takes us out of the realm of pure science and into broader issues of space policy – and that's a subject for the final chapter.

A QUESTION OF WHEN, NOT IF 8

At some point in the future, Earth will be struck by another Chicxulub-sized object – and it will cause the same kind of mass extinction that wiped out the dinosaurs. That's a certainty, not a possibility. The only question is how far in the future it will happen. Chicxulub was 66 million years ago, and the next similar event might be another 66 million years in the future. If that's the case, it's unlikely there will be any humans left to worry about it – there are plenty of other ways we could become extinct before then. At the other extreme – if a ten-kilometre comet came plummeting in from the Oort Cloud on a particularly unfortunate trajectory – we could see a Chicxulub-sized impact next year.

Don't get too worried, though – a 66-million-year wait is far more likely than a global extinction next year. That's because, as we've seen, large NEOs are far less common than small ones. So the odds favour the Earth being hit by a smaller object first. It probably wouldn't wipe out the whole species, but it could still be a serious disaster. Even if the effects were highly localised, it could destroy a large city or a small country. Another possibility is that the fragile infrastructure supporting global civilisation – the power grid, communications networks, fuel production and distribution – could be disrupted to the point where the world is thrown back into the dark ages. That's essentially the scenario depicted in Larry Niven and Jerry Pournelle's novel *Lucifer's Hammer*.

Doing something about it

We have an advantage over previous generations (and over the dinosaurs) in that we possess the technology and scientific knowledge to detect an impact ahead of time and potentially prevent it. But being able to do something isn't the same as actually doing it. The central irony of *Lucifer's Hammer* – written in 1977, and set a few years after that date – is that the capability exists but it isn't used, because no one takes the threat seriously enough. The novel does feature a group of astronauts in orbit – on a joint American-Russian space mission similar to the International Space Station – but they can only watch the disaster unfold below them. When they eventually return to Earth in their Soyuz capsule, technological civilisation is in ruins. As they sombrely observe, they're destined to be the last astronauts 'for about a thousand years'.

Unfortunately, this scenario is much more likely than the optimistic one portrayed in *Armageddon* and *Deep Impact*.

Simply understanding the science and knowing how to build the technology isn't enough. The same could be said of a whole range of things, from research stations on the Moon to colonies on Mars. They're perfectly possible, and yet they don't exist outside science fiction.

Of course, the situation might change if a real threat was discovered – one that had a high probability of wiping out civilisation. Suppose an incoming object with a Torino rating of 9 or 10 was discovered tomorrow – could we get our act together in time to save the species from extinction?

The most famous example of a space programme being put together in a hurry was project Apollo, which went from a standing start to the Moon in less than ten years. Although the first landing, Apollo 11, is remembered best, it was just one of a whole series of spaceflights. This entailed building a lot of space hardware in a short time; the original plan called for no fewer than ten Moon landings. As surprising as it might sound, one group of people had plans to hijack all that hardware for an asteroid deflection mission.

It wasn't a serious plan, of course, but an academic exercise. In 1967, when the Apollo programme was fully defined but before any mission had actually flown, a professor at the Massachusetts Institute of Technology gave his students a challenge. In those days, one of the few asteroids known to be on an Earth-crossing orbit was Icarus – a kilometre-wide rock with an aphelion in the asteroid belt and a perihelion less than 0.2 AU from the Sun (hence its name, from the mythological character who 'flew too close to the Sun'). As a theoretical exercise, the students were asked to imagine that Icarus would crash into Earth on its next close pass, just over a year later (actually there was never any chance of it hitting).

The students looked at what could be done, within the short time available, if all the Apollo hardware were repurposed for an asteroid intercept mission instead of a Moon landing. That involved ditching the human crew and replacing them with the most powerful nuclear warheads of the day. They came up with a timeline for making all the modifications and carrying out all the rocket launches which actually looked feasible. The fleet of repurposed Apollo spacecraft would fly to the asteroid and explode their bombs, one after another, about 30 metres from it. The idea wasn't to blow Icarus to pieces – though that might happen – but simply to knock it off course. In the students' estimation, they had an 85 per cent chance of reducing the effect of the collision, and a 70 per cent chance of preventing it altogether.

In retrospect, the student study was over-optimistic both with regard to the speed with which the necessary modifications could be made, and the lack of engineering problems that would be encountered in the process. It was optimistic in another way too, in assuming zero delay in obtaining the necessary funds and political commitment. That's unlikely to happen in the real world. As strange as it may seem, not everyone wants to see an asteroid deflection mission.

A two-edged sword?

It's a truism that any piece of technology that works can be weaponised – and probably will be. The thought that a

Tunguska-sized impactor could, if correctly targeted, destroy a large metropolitan area has attractions for a certain mind-set. Could a technology that was designed to divert such an object away from a collision course also be used to divert it onto one?

That's the idea behind Bill Napier's edge-of the-seat thriller *Nemesis* (1998). 'The Americans suspect that an asteroid has been clandestinely diverted onto a collision course with their country,' the Astronomer Royal informs the protagonist. The novel's focus is on the political turmoil this suspicion creates – with threats of nuclear strikes, assassinations and revolutions. In a sufficiently paranoid climate, you don't even need a real asteroid, just the hint of one.

But could asteroid deflection technology really be misused as a weapon? An early proponent of the idea that it could be was Carl Sagan, who coined the term 'deflection dilemma' to describe the situation. In reality, however, the dilemma is largely an illusion. As we saw in the previous chapter, there are millions of ways a NEO can pass close to the Earth without hitting it – and only one way that it can hit. If you want to shift the NEO from that one hit trajectory onto a miss trajectory, there's a lot of room for error. If you move it in the opposite direction to the one you intended, it's still going to miss.

The mirror-image problem is enormously harder. If you've got an asteroid that's going to miss, and you want it to hit – perhaps even to hit a precise spot on Earth, such as Washington DC – there's no room for error at all. It becomes a problem in precision targeting, not simply pushing something off course.

There's another difference, too. In order to turn a hit into a miss, you only need a deflection of the order of one Earth radius. But to turn a miss – the kind that might come up by chance in a militarily useful timeframe – into a hit, you're almost certainly going to need a much bigger deflection. Typical 'close passes', even those that make scaremongering newspaper headlines, are measured in tens or hundreds of Earth radii.

So Carl Sagan's deflection dilemma is basically a non-problem. That still leaves plenty of other ways people can object to asteroid deflection technology, though. Sagan's argument had the virtue of being a rational one – but not everyone in the world is rational. There's never any shortage of irrational arguments against any scientific topic. The same mindset that can dismiss climate change as an establishment hoax can dismiss impact events for the same reason. Not everyone will believe that an object a kilometre across can have a devastating effect on the whole planet.

Sticking to the theme of irrationality – whenever an upcoming NEO near-miss is announced, there are always plenty of conspiracy theorists ready to claim that the authorities are lying and it's actually going to hit. In practice, you don't even need a real event to trigger this sort of lunacy, and impact scares often pop up spontaneously on the internet. They can't be contradicted, either, since the whole point of the claim is that NASA, the government and the science community are conspiring to hide the truth.

If you didn't know anything about the conspiracy mentality you might assume that, if a real impact threat were to emerge, conspiracy theorists would finally accept that they were being told the truth. But that never, ever happens. Online conspiracy forums would be abuzz with the idea that it's all a hoax – and that whatever the establishment wants to do by way of a solution, they shouldn't be allowed to get away with it.

As with any big issue, feelings on both sides are likely to run at fever pitch. Benford and Rotsler's *Shiva Descending* makes exactly this point. As NASA desperately tries to mount a space mission to divert the incoming asteroid, its efforts are hindered at every step – on one side by techno-phobic anarchists, on the other by religious zealots ('The great asteroid from space is our destiny! One cannot change destiny!'). By a perverted kind of illogic, the latter – having convinced themselves that the end of the world is inevitable – go all out to sabotage NASA's attempt to avert it.

It's all too believable. There's no such thing as a big issue on which everyone in the world – or even the electorate of a single country – agrees. It's far more common to have a roughly 50-50 split between two completely opposite views. There's no reason to suppose that an asteroid deflection mission would be any different – especially if it involved any kind of nuclear technology.

This is the real 'deflection dilemma' – and something like it crops up in Arthur C. Clarke's novel *The Hammer of God*. After the 'slow-push-pull' of Plan A fails, all that's left is the nuclear Plan B. In true democratic fashion, the world's population is offered a referendum on the subject:

According to best estimates, Kali now has (1) 10 per cent probability of impacting Earth; (2) 10 per cent probability of grazing the atmosphere, causing some local damage by blast; (3) 80 per cent probability of missing Earth completely. Plans are being drawn up to detonate a thousand megaton bomb on Kali, thus splitting it into two fragments. ... On the other hand, disrupting Kali may result in the bombardment of much more extensive areas of Earth, by smaller but still highly dangerous fragments. You are accordingly asked to vote on the following proposition. The bomb should be detonated on Kali: A. Yes; B. No; C. No opinion.

One thing's for sure – not many people would tick box C.

Looking beyond Earth

Space travel is essential for the long-term survival of the human species. That's not just to deal with NEOs, but to provide an alternative to staying on Earth forever. Politicians are fond of saying 'there is no Planet B' – and there isn't, if they don't make any effort to create one. That's a thread that runs through *Shiva Descending*, which is still relevant today even though it was written in 1979. That year was a particularly depressing time in the American space programme, when it was shrinking back into Earth orbit after the last of the Apollo missions.

Benford and Rotsler set their novel a few decades after that time, in an extrapolated version of the early 21st century that looks remarkably like our own present. When the Shiva crisis arises – an asteroid on a direct collision course for Earth – a long-serving astronaut launches into a tirade against the government's chronic under-investment in space:

> If America had half the brains of a goat, it would have been building colonies years ago. Mankind wouldn't be wiped out – ever – by some fucking rock, or anything else.

The only response the NASA administrator can give is a pathetic one: 'But we do have several space stations and they are sending out a steady supply of information about the biosphere.' As it turns out, we don't even have 'several' space stations – just the one.

At the risk of spoiling the plot for people who haven't read the novel, *Shiva Descending* has an unexpectedly happy ending. Despite numerous sabotage attempts and other mishaps, the protagonists eventually succeed in diverting Shiva from its collision course. Okay, you expected that – but perhaps not the bit that comes next. Although it wasn't a deliberate part of the plan, Shiva is knocked out of its Sun-centred orbit and into one that circles the Earth. It's a big orbit, way out beyond the Moon, but it's still accessible to Apollo-type spacecraft.

This suddenly transforms a threat into an opportunity. As one of the characters puts it:

> Shiva is a vast mountain of iron and, I'm told, other valuable elements. Now they are all close to Earth. In orbit, where we can get at them. Near the orbital factories. New raw material and a fantastic amount of them. Shiva itself

can be hollowed out in the process of mining and made liveable. Space colonies ... the real thing, with a potential for economic growth and self-sufficiency.

It would be nice to think we could jump straight to that stage in the real world, without having to go through all the angst of a potential impact first. If space travel were the sole province of government, as it was when Benford and Rotsler wrote their novel, it would probably never happen. Today, though, there are other influencers in the space business – and some of them have a completely different perspective.

Take Elon Musk, for example – whose car made a cameo appearance earlier. His whole attitude to space is different from NASA's. For Musk, it's not about science or exploration – it's about long-term survival. As he said in 2017, 'One path is we stay on Earth forever, and then there will be some eventual extinction event. ... The alternative is to become a spacefaring civilisation and a multiplanetary species.' As far as Musk is concerned, this puts space travel right up there with humanity's other top priorities, like eradicating poverty or disease. On another occasion, he put it this way:

There is a strong humanitarian argument for making life multiplanetary, in order to safeguard the existence of humanity in the event that something catastrophic were to happen, in which case being poor or having a disease would be irrelevant, because humanity would be extinct. It would be like, 'Good news, the problems of poverty and disease have been solved, but the bad news is there aren't any humans left'.

Musk's current focus is on the planet Mars, but there are other places humanity could expand to. The Moon has fewer natural resources than Mars, but it's closer and easier for us to get to. Another possibility is to build huge artificial habitats in space – either orbiting the Earth or in their own independent orbits around the Sun. There are several moons of Jupiter and Saturn that might be suitable for colonisation, too.

So which of these is going to be our 'Planet B'? The best answer – to maximise humanity's chances of surviving a cosmic impact – is all of the above. The only certainty is that at some point in the future an asteroid or comet will collide with planet Earth. How we cope with it is up to us.

FURTHER READING

Chapter 1: Asteroid Apocalypse

Brian Clegg, *Armageddon Science* (St Martin's Griffin, 2012)

Andrew May, *Pseudoscience and Science Fiction* (Springer, 2017)

Carl Sagan & Ann Druyan, *Comet* (Headline, 1997)

Chapter 2: Rocks in Space

John Man, *Comets, Meteors and Asteroids* (BBC Books, 2001)

Matt Salusbury, 'Meteor Man', *Fortean Times*, August 2010,
 pp. 40–45

William J. Broad, 'Flecks of extraterrestrial dust, all over the
 roof', (https://www.nytimes.com/2017/03/10/science/
 space-dust-on-earth.html)

Alessandra Potenza, 'Track Elon Musk's Tesla Roadster in space'
 (https://www.theverge.com/2018/2/17/17019796/where-is
 -roadster-website-tesla-spacex-elon-musk-falcon-heavy)

Chapter 3: Collision Course

Kenneth Chang, 'You could actually snooze your way through an
 asteroid belt' (https://www.nytimes.com/2016/04/05/

science/you-could-actually-snooze-your-way-through-an
-asteroid-belt.html)

NASA Centre for NEO Studies, *NEO Basics* (https://cneos.jpl
.nasa.gov/about/basics.html)

David W. Hughes, 'The position of Earth at previous apparitions
of Halley's comet', *Quarterly Journal of the Royal Astronomical
Society*, December 1985, pp. 513–520

NASA, 'A gravity assist primer' (https://solarsystem.nasa.gov/
basics/primer/)

Mike Wall, 'Biggest Spacecraft to Fall Uncontrolled From Space'
(https://www.space.com/13049-6-biggest-spacecraft-falls
-space.html)

Chapter 4: Death from Space

Larry Niven & Jerry Pournelle, *Lucifer's Hammer* (Orbit, 1978)

Gerrit L. Verschuur, *Impact! The Threat of Comets and Asteroids*
(Oxford University Press, 1996)

Charles Q. Choi, 'Impact! New Moon craters are appearing faster
than thought' (https://www.space.com/34372-new-moon
-craters-appearing-faster-than-thought.html)

Jeremy Plester, 'Krakatoa – death, destruction and dust'
(https://www.theguardian.com/news/2013/aug/25/
weatherwatch-krakatoa-explosion-tsunami-disaster
-dust-sunsets)

BBC News, 'Meteorites injure hundreds in central Russia'
(http://www.bbc.co.uk/news/world-europe-21468116)

Chapter 5: Cosmic Connections

Lisa Randall, *Dark Matter and the Dinosaurs* (Vintage, 2017)

Victor Clube & Bill Napier, *The Cosmic Serpent* (Faber, 1982)

Seth Shostak, 'Is this mysterious space rock actually an alien
spaceship?' (https://www.nbcnews.com/mach/science/
mysterious-space-rock-actually-alien-spaceship
-ncna829501)

Edward J. Steele et al. 'Cause of Cambrian Explosion: Terrestrial or Cosmic?' (https://www.sciencedirect.com/science/article/pii/S0079610718300798)

Chapter 6: Mapping the Threat

Carrie Nugent, *Asteroid Hunters* (TED Books, 2017)

Bill Napier, *Nemesis* (Headline Feature, 1998)

Brian Clegg, 'Schedules in Space', *Impact*, issue 4, pp. 6–9 (https://www.theorsociety.com/Pages/Impact/issue4.aspx)

Sarah Knapton, 'Alien life unlikely on Rosetta comet, say mission scientists' (https://www.telegraph.co.uk/news/science/space/11720871/Alien-life-unlikely-on-Rosetta-comet-say-mission-scientists.html)

Chloe Cornish, 'Interplanetary players: a who's who of space mining' (https://www.ft.com/content/fb420788-72d1-11e7-93ff-99f383b09ff9)

BBC News, 'Invest to avert Armageddon' (http://news.bbc.co.uk/1/hi/uk_politics/289733.stm)

Defending Planet Earth: Near-Earth object surveys and hazard mitigation strategies (https://www.nap.edu/read/12842/chapter/1)

Chapter 7: Planetary Defence

Arthur C. Clarke, *The Hammer of God* (Orbit, 1994)

Gregory Benford & William Rotsler, *Shiva Descending* (Sphere Books, 1980)

Lawrence Livermore National Laboratory, *Proceedings of the Planetary Defence Workshop* (https://e-reports-ext.llnl.gov/pdf/232015.pdf)

Brent W. Barbee et al., 'Options and uncertainties in planetary defence: Mission planning and vehicle design for flexible response' (https://esc.gsfc.nasa.gov/media/491 – includes PDF of *Acta Astronautica* article)

NASA Goddard Media Studios, 'How Sunlight Pushes Asteroids' (https://svs.gsfc.nasa.gov/11964)

Chapter 8: A Question of When, not If

David Portree, 'MIT saves the world: Project Icarus'
(https://www.wired.com/2012/03/mit-saves-the-world
-project-icarus-1967/)

Russell L. Schweickart, 'The real deflection dilemma'
(http://research.dynamicpatterns.com/wp-content/uploads/
2011/03/The-Real-Deflection-Dilemma-Schweickart.pdf)

Ross Andersen, 'Elon Musk puts his case for a multi-planet
civilisation' (https://aeon.co/essays/elon-musk-puts-his
-case-for-a-multi-planet-civilisation)

Elon Musk, 'Making humans a multiplanetary species'
(https://www.liebertpub.com/doi/full/10.1089/
space.2017.29009.emu)

INDEX

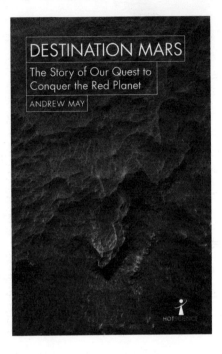

When the Apollo astronauts walked on the Moon in 1969, many people imagined Mars would be next. Half a century later, only robots have been to the Red Planet and our astronauts rarely venture beyond Earth orbit.

Now, Mars is back. With everyone from Elon Musk to Ridley Scott and Donald Trump talking about it, interplanetary exploration is back on the agenda and Mars is once again the prime destination for future human expansion and colonisation.

In *Destination Mars*, astrophysicist and science writer Andrew May traces the history of our fascination with the Red Planet and explores the science upon which a crewed Mars mission would be based, from assembling a spacecraft in Earth orbit to surviving solar storms. With expert insight he analyses the new space race and assesses what the future holds for human life on Mars.

ISBN 9781785782251 (paperback) / 9781785782268 (ebook)